COMPUTER NETWORKS IN
THE CHEMICAL LABORATORY

EDITED BY
GEORGE C. LEVY
DAN TERPSTRA
DEPARTMENT OF CHEMISTRY
THE FLORIDA STATE UNIVERSITY
TALLAHASSEE, FLORIDA

COMPUTER NETWORKS IN THE CHEMICAL LABORATORY

Presented in part as a Symposium
at the 179th National Meeting of the
American Chemical Society,
Houston, Texas, March, 1980.

A WILEY-INTERSCIENCE PUBLICATION
JOHN WILEY & SONS, New York • Chichester • Brisbane • Toronto

Copyright © 1981 by John Wiley & Sons, Inc.

All rights reserved. Published simultaneously in Canada.

Reproduction or translation of any part of this work beyond that permitted by Sections 107 or 108 of the 1976 United States Copyright Act without the permission of the copyright owner is unlawful. Requests for permission or further information should be addressed to the Permissions Department, John Wiley & Sons, Inc.

Library of Congress Cataloging in Publication Data:

Main entry under title:
   Computer networks in the chemical laboratory.

   Includes index.
   1. Chemistry—Data Processing—Congresses.
2. Computer networks—Congresses. I. Levy, George C.
II. Terpstra, Dan. III. American Chemical Society.

QD39.3.E46C644   542'.8   81-599
ISBN 0-471-08471-9   AACR2

Printed in the United States of America

10 9 8 7 6 5 4 3 2 1

ERRATUM

The name of Michael K. Starling, the senior author, was inadvertently omitted from Chapter 3. (This publishing error was not the fault of the co-authors.)

His name and address should also be added to the list of contributors:

M. K. Starling
Union Carbide Technical Center
South Charleston, West Virginia

". . . No man is an *Iland,* intire of it selfe; every man is a peece of the *Continent,*

—John Donne, 1623

CONTRIBUTORS

D. W. ALDERMAN
Department of Chemistry, University of Utah, Salt Lake City, Utah

H. Ch. BROECKER
Department of Chemistry, University of Hamburg, West Germany

R. F. COLEY
Commonwealth Edison, Technical Center, Maywood, Illinois

R. E. DESSY
Department of Chemistry, Virginia Polytechnic Institute and State University, Blacksburg, Virginia

C. L. DUMOULIN
Department of Chemistry, Florida State University, Tallahassee, Florida

R. P. GOEHNER
General Electric Company, Corporate Research and Development Center, Schenectady, New York

D. M. GRANT
Department of Chemistry, University of Utah, Salt Lake City, Utah

W. D. HAMILL, JR.
Department of Chemistry, University of Utah, Salt Lake City, Utah

W. T. HATFIELD
General Electric Company, Corporate Research and Development Center, Schenectady, New York

D. J. HOOLEY
Department of Chemistry, Virginia Polytechnic Institute and State University, Blacksburg, Virginia

G. C. LEVY
Department of Chemistry, Florida State University, Tallahassee, Florida

E. LIFSHIN
General Electric Company, Corporate Research and Development Center, Schenectady, New York

C. L. MAYNE
Department of Chemistry, University of Utah, Salt Lake City, Utah

H. G. W. MÜLLER
Department of Chemistry, University of Hamburg, West Germany

C. N. REILLEY
Kenan Laboratories of Chemistry, University of North Carolina, Chapel Hill, North Carolina

D. TERPSTRA
Department of Chemistry, Florida State University, Tallahassee, Florida

W. S. WOODWARD
Kenan Laboratories of Chemistry, University of North Carolina, Chapel Hill, North Carolina

K. ZIETLOW
Department of Chemistry, University of Hamburg, West Germany

PREFACE

NETWORKING—it's one of the hottest buzzwords in the computer industry today. One can pick up almost any current issue of the popular trade magazines and find references to computer networks. Home computer network connections are now available in the United States. Consumer networks are in operation in Britain, France and other countries. New network products are being introduced almost daily, and major integrated circuit manufacturers are busily designing network interfaces on single silicon chips. International standards committees are working feverishly to come to agreement on network protocols to permit product interchangeability.

But in spite of all this excitement, the average chemist still knows little or nothing about computer networks and how they can best be employed in chemical laboratories. *Computer Networks in the Chemical Laboratory* helps to fill this informational void by providing detailed descriptions of real laboratory computer networks. The authors are, for the most part, practicing chemists who have anticipated the importance of networking in the laboratory. Their pioneering efforts are presented here to provide insight into the problems, options and opportunities of computer networking and the potential impact of networks on laboratory experimental control, data collection and management.

The first chapter of this book gives a brief historical introduction to modern laboratory computers and computer networks. The second offers a look at one of the pioneering efforts in distributing microcomputer technology in the laboratory. Chapter 3 addresses some of the software considerations of networking, and the ways in which one software language (FORTH) deals with the network environment. The computer network described in chapter 4 illustrates the power and flexibility afforded by networking in a large industrial research laboratory. Chapters 5, 6 and 7 discuss networks with varying levels of hardware and software sophistication in academic laboratory environments; and chapter 8 ends the book with a description of the implementation of a commercially available network facility (DECNET) in an industrial laboratory.

The decade of the 80's will surely see a tremendous growth in the development and application of computer networking in the laboratory as well as the office and the home. It is hoped that this volume, in its accounts of some early networking attempts, will provide insights into the possibilities of computer networks in the laboratory.

The editors would like to thank the American Chemical Society Divisions of Computers in Chemistry and Analytical Chemistry for their encouragement in organizing a symposium on this topic presented at the 179th National meeting of the Society in March of 1980. This book is an outgrowth of the information discussed at that symposium. Thanks are due also to the contributors to this book for their cooperation and patience in preparation of the necessary photo-ready text and figures. The editors as authors would also like to thank Steve Leukanech and Dick Roche for the skill and effort amply represented by the artwork in chapters 1 and 5. Finally, our thanks are offered to Leslie Heinz for her willingness to explore the exciting and occasionally frustrating new terrain of computer-based word processing in the preparation of several portions of this book.

GEORGE C. LEVY
DAN TERPSTRA

Tallahassee, Florida
January, 1981

CONTENTS

CHAPTER 1	**Introduction to Laboratory Computer Networks** **D. TERPSTRA and G. C. LEVY**	**1**
	Introduction	1
	Mainframe Computers	2
	Minicomputers	6
	Microcomputers	8
	Computer Networks	11
	Long Haul Networks. Local Area Networks.	
	Summary	22
CHAPTER 2	**A General Purpose Microcomputer for Laboratory Automation in a Hierarchial Environment** **W. STEPHEN WOODWARD and CHARLES N. REILLEY**	**25**
	Introduction	25
	The Hierarchial Scheme	25

Design Features of the General Purpose Microcomputer	27
Hardware Components	31
Central Processing Components. Real Time Data Acquisition and Experiment Control. Operator Interaction and Data Presentation Peripherals. Magnetic Media Storage. Communication with Shared-Resource Computing Hardware.	
Software Development Techniques and Resources	36
Applications	39
Documentation	40

CHAPTER 3 FORTH, A Fourth Generation Language System Implements a Network
R. E. DESSY and DAVID J. HOOLEY — 41

Background	41
FORTH, A Language, A System, A Network	43
Computer Languages, Four Generations	43
FORTH as an Interactive Compiler	45
FORTH as a Data Structure	45
FORTH as a Virtual Computer Emulator	49
A User's Introduction	51
Stack Operators	51
Mathematics	53
Variables and Storage Operators	55
Definition Types	57
FORTH as a Structured Language	58
Control Structures	58
Mass Storage and Virtual Memory	60
Relocation, Overlays, Jump Tables, Aliases and Strings	62
Multiusers	63
Multitasking	63
Dummy Tasks	66
Networking	67
Record Management and Report Generation	68
An Applications Oriented Language	69

CHAPTER 4	The General Electric Laboratory Automation System W. T. HATFIELD, R. P. GOEHNER and E. LIFSHIN	71

Introduction	71
Background	72
System Requirements	74
Configuration	76
Instruments Interfaced	79
System Interconnection	81
Operating System Software	82
Applications Software	82
Conclusion	100

CHAPTER 5	WARPATH, A Computer Network for a Multi-user Laboratory Environment G. C. LEVY, D. TERPSTRA and C. L. DUMOULIN	102

Introduction	102
The WARPATH Local Area Network	103
General Considerations. WARPATH: The Physical Layer. WARPATH Topology. The WARPATH Chief. The Intelligent WARPATH Interface. WARPATH Network Protocols.	
The "INDIANS" on the WARPATH	118
Overview. INDIAN/Spectrometer Interface. INDIAN/Operator Interface. INDIAN/WARPATH Interface.	
The Data Processing Station	145
Processing Station Overview: Hardware. Processing Station Overview: Software. FSUNMR and the User. Sample Capabilities. FSUNMR and the WARPATH Environment.	
Conclusions and Acknowledgements	158

CHAPTER 6 Hardware and Software Aspects of the Development of the Satellite Data Processing Network at the University of Hamburg
K. ZIETLOW, H. Ch. BROECKER and H. G. W. MÜLLER — 161

Introduction — 161
Hardware of the Satellite Computer Network — 162
Communication Lines — 167
Communication Software and User Programs, — 169
  Communication Line Protocols. Operating Systems and User Programs.
Experiences and Conclusions — 174

CHAPTER 7 The University of Utah NMR Laboratory Minicomputer Network
D. W. ALDERMAN, W. D. HAMILL, JR., C. L. MAYNE and D. M. GRANT — 178

Introduction — 178
Network Components — 178
Network Function — 180
Network Implementation — 182
  Spectrometer to PDP 11/04 Serial Lines. PDP 11/04 to PDP 11/70 Parallel Link. Network Programs. Plot Facility Program.
Overview — 186

CHAPTER 8 A Production Laboratory Implementation of DECNET
R. F. COLEY — 188

Introduction — 188
  Analytical Facilities and Staff.
The Time for Action — 189
  Modernization of Operating Stations. Equipping of New Stations. Developing a Comprehensive Quality Control Program.
Definition of Needs — 190
  Identified Concerns.

The Action Plan	191
Commonwealth Edison Company Analytical Instrumentation Project	192
Overview. Objectives. Project Requirements. Network Topology and Communications Hardware. Purchased Hardware and Software. Specially Designed System Software. System Operation. General Features. Special Operating Conditions. Manual Analyses. Automated Analytical Instrumentation System (AAIS).	
Implementation	203
1980. 1981–1983.	
Conclusions	211
Index	215

COMPUTER NETWORKS IN
THE CHEMICAL LABORATORY

1

INTRODUCTION TO LABORATORY COMPUTER NETWORKS

Dan Terpstra, George C. Levy

INTRODUCTION

   Chemistry, like many other experimental sciences, is becoming inundated by computers and computerized instrumentation. Although chemists are generally quick to take advantage of computers in their research, the concepts of distributed processing and computer networking are still foreign to most chemistry laboratories. However, laboratory mini- and microcomputers have become so prolific in modern research environments that a number of chemists and other scientists have begun to realize the necessity of enabling their computers to communicate with one another -- sharing data, programs, and physical resources.
   Much of this book is devoted to documenting (through first-hand accounts) the pioneering efforts of some chemists who have anticipated the importance of computer networking in the chemical laboratory. The applications presented here vary widely in scope and level of sophistication. At one end of the spectrum we see essentially autonomous minicomputers loosely connected via slow-speed serial data links. The other end is occupied by highly interdependent mini- and micro- computers tightly coupled through high-speed parallel or serial data links. The networking schemes range from commercial implementations covering a large portion of the state of Illinois to entirely "in-house" hardware and software configurations spanning a single building or laboratory. These diverse approaches are woven together with a common thread: the recognition of the emergence of digital techniqes for laboratory measurement and control, and the necessity of providing efficient methods for handling the vast quantities of information produced by such methods.
   Before dealing with specific network implementations, it is useful to construct a framework within which computer networks can be discussed. The goal of the remainder of this chapter is to provide just such a framework. A good starting point is a brief review of the historical development of laboratory

computers describing mainframe systems, the introduction of laboratory minicomputers, and the current massive implementation of microprocessors and microcomputers[1]. With this information as background, the stage is set for a discussion of computer networks including their history, common network structures, and the important distinction between so called "long-haul" and "local-area" networks. Such an overview is not designed to be comprehensive, but to acquaint the reader with some of the vocabulary and concepts employed in the remaining chapters of this book.

MAINFRAME COMPUTERS

The atomic bomb and its mixed bag of technological implications was not the only offspring of weapons research during the second World War to have a profound effect on the last half of the 20th century. Weapons research also served as the progenitor of electronic digital computers, the consequences of which are only now beginning to be felt by many sectors of society.

A major computational problem during World War II was the rapid calculation of projectile trajectories for newly developed weapons. One mathematician estimated[2] that it would require 10 to 20 years for a person operating a mechanical calculator of the day to calculate a complete firing table for one such new weapon. Various mechanical and electro-mechanical devices were employed to actually perform these calculations. A relay computer called the Model III by Bell Telephone Laboratories (that could calculate such a table in roughly two months) was completed in 1944. It used 9000 relays, weighed approximately 10 tons, and could multiply two 7 digit decimal numbers in about a second.

The idea of a completely electronic calculational machine employing vacuum tube technology was introduced by Prof. J. W. Mauchly of the University of Pennsylvania's Moore School of Electrical Engineering in 1942. Funded by the government in 1943, the University developed the Electronic Numerical Integrator and Computer. ENIAC, as it was called, was too late for the war effort with its introduction in February of 1946, but it was used by the Government at both Los Alamos and the Aberdeen Proving Ground until 1955. This first fully electronic computer was immense by today's standards. It required 18,000 vacuum tubes and 150 kilowatts of electrical power to run. It occupied 1500 square feet and weighed 30 tons. The equivalent in computing power today would fit on a silicon chip the size of a fingernail and consume less than a watt of power. ENIAC could, however, perform a 10 digit decimal multiplication in less than 3 milliseconds -- about 3 orders of magnitude faster than the Bell

Model III introduced only two years earlier. ENIAC's major flaw lay in the fact that programming was done by physically plugging wires into a patch panel, a task that could take even an experienced operator many days to complete. This major drawback spawned computers in which the program, like the data, is maintained in memory -- as is done on modern computers. Two of these computers were the EDSAC (Electronic Delay Storage Automatic Computer) built at Cambridge University in 1949, and EDVAC (Electronic Discrete Variable Computer) introduced by the Moore School in 1950.

The solid state era was heralded in late 1947 with the invention of the point-contact transistor by William Shockley, John Bardeen, and Walter H. Brattain of Bell Telephone Laboratories. Although they received the Nobel Prize in 1956 for their efforts, the initial reception of their device was quite cool. In fact, transistors weren't incorporated into computers until the late 1950s when they were employed both by IBM and Univac. By this time, magnetic core memories had already replaced other forms of computer memories, improving speed by another factor of 100 over earlier models. Both magnetic tape and the newly introduced magnetic disks were now available as mass storage devices. User interaction on these early computers was primarily through the use of punched cards and line printers, a legacy that has remained with us for far too long.

Software development was originally much more slowly paced than hardware development, since the concept of a "program" itself was still crystallizing. Univac was an early leader in making computers easier to program. In the early 1950s, virtually all programming was done in machine language -- binary strings of 1s and 0s, perfect for the computer but almost indecipherable for the human. In an effort to alleviate this problem, Univac introduced the first interpreter in 1952. It read assembly code mnemonics line by line and interpreted them into machine language commands that were then immediately executed. Univac also introduced the first compiler language at about this time in which a program written in algebraic notation was compiled into a machine language program at one time by the computer. This machine language version could then be directly executed by the computer at any later time. IBM made a contribution to compiled computer languages in 1957 with the introduction of its Formula Translation programming language, otherwise known as FORTRAN. In 1959, two more popular computer languages were introduced: ALGOL (Algorithmic Language) -- developed jointly by the Association for Computing Machinery and the German Association of Applied Mathematics, and COBOL (Common Business Oriented Language) -- an offering from the U.S. Pentagon.

As mainframe computers entered the 1960s, programming became more and more sophisticated. The operating system, or "program to run programs" was introduced early in the decade. This eliminated the need for the user to be concerned about most of the "housekeeping" chores associated with running a program. Also in the 60s, timesharing developed as a way to maximize hardware resources. Before this time, most computers executed programs sequentially in what is known as "batch" execution. Dartmouth College and General Electric pioneered the implementation of a system in which computer resources were rapidly shared between many users and many programs, giving each one a small "share" of the computer's time (hence the name) before moving on to the next user and program. One of the results of timesharing at Dartmouth was to make the computer more accessible to large numbers of students. This led to the need for an easy to learn interactive computer language for instructional purposes. Such a language was BASIC (Beginner's All-Purpose Symbolic Instruction Code), developed at Dartmouth in 1964.

Hardware improvements were also far-reaching in the 60s. The integrated circuit was first demonstrated by Texas Instruments in 1958, and was becoming incorporated into a third generation of computers by 1964, supplanting the first-generation vacuum tube and second-generation discrete transistor based machines. Main memory was still prohibitively expensive, so "virtual memory" techniques were developed to move into mass memory the programs or data that were not currently in use and recall them as they were needed. Multiprogramming techniques and parallel processing, in which many streams of data were simultaneously operated upon, were both pioneered during this period. One parallel processor, the Illiac IV, was begun in 1967 by Burroughs Corporation. It contained 64 parallel processing elements, a megabyte of main memory, and could execute instructions at the rate of 200 million per second. Life was made easier for the user of large systems with the gradual replacement of teletypes and card readers by cathode ray terminals (CRTs). And the interchange of alphanumeric information was simplified by the introduction in 1963 of the American Standard Code for Information Interchange (ASCII) which is in essentially universal use in computer peripherals today.

By the 1970s the mainframe computer arena had stabilized in most instances. One writer states, "Instead of the revolutionary changes in fundamental organization or architecture that occurred in the 1950s and 1960s, developments in the 1970s were more evolutionary."[3] This was primarily due to the fact that too much money and time had been invested in hardware specific programs to make affordable any dramatic changes in the architecture on which those programs had to run. Integrated circuits were gradually increasing in complexity, however, with the result that earlier architectures could be packed into much smaller boxes with much smaller price tags and power requirements, and concomitantly with higher speeds. In addition, semiconductor memory technologies had matured to the point where main memory limitations were no longer as constricting as they once had been. Large Scale Integrated (LSI) circuits on single silicon chips made it possible to disperse some of the intelligence of the computer into its peripheral devices, thus easing the overhead of monitoring many functions at once. This carried the concept of timesharing one step further into the realm of distributed processing and data communications networking, an area that will see extensive development in the current decade.

There were some dramatic mainframe architectural advances in the 70s. One of the most spectacular was (and still is) the Cray-1 supercomputer, a vector processor introduced in 1976 by Seymour Cray, a founder of and former designer for Control Data Corporation (CDC). The computer itself stands roughly seven feet high in a cylindrical configuration to minimize interconnection distances, and is liquid-cooled to maintain proper operating temperatures. It can execute between 80 and 130 million floating point instructions per second, giving it current claim to the position of world's fastest computer. This claim may be challenged in the near future when Lawrence Livermore Laboratory introduces its new S-1 Mark IIA supercomputer, said to perform up to 400 million floating point operations (400 megaflops) per second. In addition, CDC has already announced its Cyber 205, claiming a maximum speed of up to 800 megaflops per second. Not one to rest on its laurels, Cray Research is reported to be working on the Cray-2, sure to keep it in competition for the world speed title[4].

The future for mainframe systems is difficult to predict, particularly since increasing levels of integration are making it virtually impossible to delineate the divisions between mainframe, mini- and micro-computers. Continuing trends in computer networking and distributed intelligence will inevitably obscure the boundaries of the "mainframe" computer as a monolithic device. It will become increasingly difficult to distinguish at what point one computer "ends" and the next "begins". This trend toward decentralization of computational hardware may find itself reversed to some extent by the impending implementation of Josephson junction devices. These devices offer nanosecond switching times and extremely low power consumption[5]. An entire computer using such devices could be built in eight cubic inches and operate at a factor of 5 faster than even the new supercomputers. One major drawback of Josephson junction technology is that this eight cubic inch computer is superconducting, and thus must operate at liquid helium temperatures, requiring complex support facilities and reinforcing the centralization of computational power. As unpredictable as the future of mainframe computers may be, they will certainly continue to play an active role in the computerization of society.

MINICOMPUTERS

Even as mainframe computers were becoming larger, faster and more sophisticated in their move from the transistors of the second generation to the integrated circuits of the third generation, another computer revolution was beginning that would soon profoundly affect scientific research laboratories. The first rumblings of this revolution were heard as early as 1959 when the newly formed Digital Equipment Corporation (DEC) began marketing its Programmed Data Processor, the PDP-1. This 18-bit computer was inexpensive for computers of that day, selling for an average price of $120,000. It took six years and seven models before DEC introduced what was to become the most successful minicomputer in history: the PDP-8, introduced in 1965. This computer had a capacity of up to 4096 12-bit words of core memory and sold for less than $20,000. It was the first mass produced computer, and the first to be given the name "minicomputer". The PDP-8 was not alone for long. By 1966, Hewlett-Packard had marketed its first "instrumentation computer" and in 1967 introduced a 16-bit general purpose mini-computer as well. Data General was founded in 1968 by three former DEC employees and introduced its 16-bit NOVA line that same year at a base price of only $8000. By this time the minicomputer was firmly established in industrial and laboratory environments nationwide.

The PDP-8 and its younger siblings and rivals ushered in an entirely different approach to the marketing and application of computers. Since the devices were mass-produced, mass-marketed, and highly competitive, the profit margin was much less than on mainframe systems. Minicomputer companies could not afford and did not provide the extensive hand-holding and customer support that had become standard with mainframes. Small profit margins also inhibited the development and sale of higher level language facilities since the user could rarely justify the high relative cost of such additional software capabilities.

The cost of software was not the only reason for a primary reliance on assembly programming. Minicomputers often were simply not big enough to support high level compilers or interpreters. Expensive main memory and severe memory address limitations demanded space efficiency that could often only be achieved through assembly level programming.

The new minicomputers were designed for the real world. They generally had short word lengths and high speeds, ideal for many real-time laboratory control and data acquisition requirements. Much attention was devoted to their input and output structures, making them easier to interface to laboratory equipment. They were generally designed to be single-user, single-task machines, often equipped with analog-to-digital and digital-to-analog converters. The applications to which these computers were put often demanded high program execution speeds, another strong argument at the time in favor of direct assembly language programming. Since the computers themselves were so inexpensive, they were often used in situations where budgets would not allow for sophisticated user input and output. The most commonly used I/O devices were the paper tape reader/punch and the teletype.

Minicomputers matured during the early 70s and as they did, they found new applications that were previously non-existent. The word lengths of these laboratory computers varied from the original 12-bit PDP-8 to the 20-bit Nicolet 1080, but most minicomputers had settled on 16-bit words as the optimum cost-effective architecture. In chemistry laboratories, minicomputers became attached to x-ray crystallographic equipment, to esr and mass spectrometers, and to gas and liquid chromatographic equipment. In some cases, the availability of the computer changed the entire texture of the experiment. For example, the implementation of the Fast Fourier Transform algorithm on minicomputers created fundamentally different methods of measurement for nmr and infrared spectroscopy.

The power of LSI circuitry resulted in greater sophistication for minicomputers in the late 70s. Large minicomputers began to look very similar to their mainframe counterparts. As minicomputers penetrated the small business and word processing environments, they developed memory remapping techniques to circumvent limited memory address space, they began to understand high level languages, and they supported increasingly sophisticated input and output devices. In addition, it was no longer necessary to restrict minicomputers to single users or single processes. Minicomputer operating systems became more intelligent and incorporated more of the features and facilities of mainframe operating systems. Minicomputer hardware incorporated priority interrupts, direct memory access, and other features that allowed the monitoring and control of many simultaneous real-time events.

By the end of the decade, some minicomputers had grown to 32-bit word lengths and the distinction between low-end mainframe and high-end mini became more a matter of semantics than of substance. At the same time, 16-bit microcomputers were encroaching on the lower end of the minicomputer spectrum, making the minicomputer's hold on its niche look precarious indeed. In fact, by 1980, a number of microcomputers outstripped the capabilities of their bigger brothers. The role of instrumental control that once belonged solely to the minicomputer is now often being given to a handful of microcomputers with the result that the minicomputer may soon become the "laboratory manager" for a number of intelligent instruments tied together by a local-area network. This course of events is already occurring, and is in fact one of the driving forces behind the accounts in this book.

MICROCOMPUTERS

The microcomputer, or microprocessor as the central processing chip itself is often called, had its beginnings in the electronic desktop calculators of the late 1960s. LSI circuits were becoming increasingly complex and calculator manufacturers were rapidly incorporating them into their products. Late in 1969, Intel Corporation was asked by a Japanese calculator firm to design an LSI chip for their calculators. Intel designed a chip that was flexible enough to be considered a general purpose computer as well as a calculator chip, and the microprocessor era was born. This chip, called the 4004, was announced in the middle of 1971. It and its three support chips comprised all of the elements of a simple computer. The 4004 cpu chip could handle roughly eight thousand bytes (8 bits) of memory and had a four bit word length. Also in 1969, both Intel and Texas Instruments were approached to design a slightly more complicated chip for an intelligent computer terminal. Intel succeeded, but its design was more powerful and much slower than what was required. Texas Instruments received the contract and Intel was left with the design for an eight bit version of its 4004. Early in 1972 Intel began marketing this device as the world's first eight bit microprocessor, the 8008. The 8008 initially sold for $200 but required ten times its worth in peripheral circuitry to turn it from microprocessor into microcomputer. Many other companies began marketing microprocessors and microcomputers, and by 1975 there were more than forty from which to choose.

Microprocessor applications soon stratified into two distinct areas. The first included control functions and rudimentary programming that were best served by the simplest of devices such as the 4004 or the popular TMS 1000 from Texas Instruments. These applications were as diverse as washing machines, microwave ovens, automobile carburation, and the like. Advances in this area tended to put increasing amounts of sophistication into a single silicon chip until by the late 70s chips were available that could, for example, convert an analog input into digital form, manipulate it, and convert it back into analog as an output. Single chips also became available that contained all the elements of a computer, including the processing unit, program memory, data memory, and input/output structures.

The second area of microcomputer applications emphasized the computer-like features of these amazing devices, and before long threw them into direct competition with minicomputers. As the speed and power of microprocessors improved, they began to be employed as actual computers.

In 1974, Intel introduced what was to become the most sought after 8-bit microprocessor ever: the 8080. It rapidly found its way into a wide variety of applications, and in the process spawned a competing company (Zilog) with a superior product: the Z-80, a faster 8-bit microprocessor designed to run programs written for the 8080, as well as programs using the expanded Z-80 instruction set. Intel combatted the Z-80 with an upgraded version of the 8080 that was called the 8085 and eventually ran even faster than the 4 MHz Z-80A. The Intel family of LSI chips was marketed by Intel both at the chip level, and at the board or system level. Intel's Intellec systems utilized the Multibus interconnection system, which has recently become standardized by the Institute of Electrical and Electronics Engineers as IEEE-796.

Another important development that followed on the heels of the 8080 was the S-100 bus. Originally marketed by Altair in 1976, the S-100 was intended as a vehicle for introducing the 8080 to computer hobbyists. A number of small companies began selling low priced S-100 bus compatible hardware, and the S-100 rapidly became a de facto standard in small business and research applications as well as in computer hobbyist circles. The IEEE recognized the popularity of the S-100 bus, and in 1979 it proposed an industry-wide standard (IEEE-696) for S-100 compatible products. In just a few short years, microprocessor chips had become not just smart controllers to be hidden inside a product, but computers in their own right, able to compete head-to-head with minicomputers.

High performance 16-bit microcomputers arrived on the scene late in 1978 with the 8086 from Intel, followed by Zilog's Z8000, Motorola's 68000, and National's 16000. Support circuitry on single silicon chips began to proliferate for microcomputers. It is now possible, for example, to obtain single package devices that can control floppy disk drives, handle sophisticated data encryption schemes, do floating point arithmetic, or even interface to local-area computer networks. In fact, since each of these devices contains a microprocessor of its own, it is not at all uncommon to find a microcomputer system that contains half a dozen or more microprocessors operating in parallel under the control of a master processor to coordinate its computational tasks.

From the outset, the software tools available for microcomputers were much more extensive than those for minicomputers at an earlier, parallel stage of their development. There are a number of reasons for this, not the least of which is the dramatic and continuing reduction in cost of semiconductor memory. Most 8-bit microcomputers are designed to support 64 kilobytes (one byte equals eight bits) of memory, and many microsystems are already finding this limit restrictive. This would have been unthinkable a decade earlier when minicomputers were being designed to support only four to eight kilowords of memory. The newer 16-bit microcomputers can handle megawords of memory, still too expensive for most implementations, but indicative of further projected reductions in memory costs.

Another reason for the plethora of excellent microcomputer software tools is the development of cheap and reliable mass storage in the form of floppy disks. A floppy disk is an eight inch or five and one quarter inch flexible plastic disk coated with iron oxide and kept in a padded folder for protection. It looks much like a 45 rpm record, but it can hold up to half a megabyte or more of programs and data for less than five dollars. The first comprehensive floppy disk operating system designed specifically for microcomputers was, like so many other microcomputer firsts, developed in conjunction with the Intel 8080. This operating system, CP/M, was introduced in 1975 by a company that has since become Digital Research[6]. CP/M provides facilities formerly available only on mini- and mainframe computers. Much in the same way that the S-100 bus became a hardware standard for the 8080 family, CP/M is becoming a software standard. Digital Research has since marketed the first multi-user microcomputer operating system (MP/M) and the first network microcomputer operating system (CP/Net); both are essentially supersets of CP/M and both run on the 8080 family of microcomputers.

Large main memory sizes, readily available mass memory, and the huge numbers of microcomputers sold, all coupled with the experience gained in the development of minicomputers, made it profitable to develop high level software for microcomputers. The first resident high level language for micros was PL/M, a modification of IBM's PL/1 language introduced by Intel for the 8080 in 1974. Today a subset of PL/1 itself is available under the CP/M operating system, along with a wide variety of BASIC interpreters, Fortran, Cobol, Pascal, or any one of a number of other languages. In addition to this broad spectrum of languages, excellent hardware and software debugging tools are available in the form of emulators and supervisory programs to speed software testing.

As microprocessors get more powerful and more specialized, it seems inevitable that the tasks to be performed will be fragmented into a number of common "building block" sub-tasks, each handled by its own dedicated microprocessor. Building a computer will then be little more than deciding how many and what types of these "building blocks" are needed to fulfill a given set of requirements, and connecting the blocks in the manner prescribed by the makers of the circuits. This leads directly to the idea of distributed or multiprocessing where a number of processors work together, each fulfilling a small part of the total function of the computer. Such schemes are becoming increasingly visible as we enter the decade of the 80s and, as should be obvious from the nature of this book, are already underway in some contexts. They will be discussed in futher detail in the remainder of this chapter.

COMPUTER NETWORKS

In its simplest and most general definition, a computer network is the means by which discrete computers exchange electronic information. However, as is often the case, this simple definition harbors a wealth of variation and complexity. For example: the number of computers involved in a given network may be as few as two or as many as several hundred or even several thousand; the computers in a network may all be identical, sharing machine level programs and memory resources, or they may be totally dissimilar, sharing only data and mutually defined messages; a computer network may cover as little as a few meters via simple twisted-pair or coaxial cable, or it may span the globe using communications satellites and microwave transmission techniques.

Despite the number of apparent differences, all computer networks share certain common features. These features have been identified and broken into seven somewhat arbitrary "layers" (illustrated in Figure 1.1) by national and international committees concerned with such matters[7,8]. All seven of these layers need not be present in every network implementation, but taken together they provide a comprehensive architecture for computer networking. The sole purpose of any given layer in this network heirarchy is to communicate with the same layer in another network computer. This is referred to as peer-to-peer communication. In order to fulfill this function, a layer must be able to pass information to and from the layers immediately above and below it (called layer-to-layer communication). Thus, for example, the network layer in one computer would communicate with the network layer in another computer by passing information down to the link layer and finally to the physical layer where peer-to-peer communication transfers the information to the other computer. It would then be passed back up through the physical and link layers of the second computer and received as peer-to-peer communication by the network layer of the second computer. To complete this transfer, the link layer, for example, is only required to pass information between itself and the network layer above it, or between itself and the physical layer below it. With this scheme, the structure of any individual layer can be modified independently of other layers, as long as the interface between layers is kept constant.

In order to clarify the roles played by each of these seven network layers, an analogy may prove helpful:

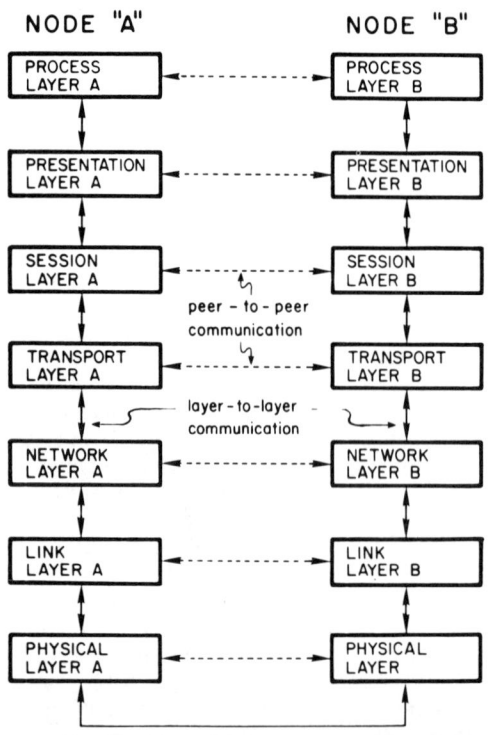

FIGURE 1.1. The seven layers of network protocol between network node A and network node B.[8]

<u>Process Layer</u>: Distinguished Professor A has conceptualized a brilliant new hypothesis that she wishes to share with her colleague, Doctor B (a peer-to-peer communication). She realizes that she cannot share her mental processes directly with Dr. B, but must convert them to a presentable verbal or mathematical form. In much the same way, specific user input or program results (ideas) form the process layer for a given computer, but are not directly transferrable to another computer.

<u>Presentation Layer</u>: Prof. A composes her hypothesis into a collection of sentences and formulae that she feels she can present to Dr. B (process layer-to-presentation layer interface). The computer must also convert specific data into a format understandable by another computer.

<u>Session Layer</u>: Prof. A decides that she will discuss her hypothesis in a verbal conversation with Dr. B. Implicit in this decision is a set of social amenities that will serve as cues for the beginning, middle, and end of the conversation session: "Hello Bill, this is..."; "I called about..."; "Give my best to

Sally and the kids..."; etc. Session protocols in a network context also define the beginnings and ends of messages, and the ways in which information is exchanged by communicating computers. Often message meanings are position dependent; if not, other cues must be provided to make the message meaningful.

<u>Transport Layer</u>: Prof. A informs her secretary that she wishes to talk with Dr. B. It is then the secretary's responsibility to contact Dr. B's secretary and establish the means by which Prof. A's message can be transported (the telephone link). A network-independent program in the sending computer has the same role as the secretary in establishing contact through the transport medium with the transport layer of the receiving computer. Neither the secretary nor the transport layer is required to know anything about the mechanism of transport except how to interface to it. This is important in that it points out that the top four layers of this definition are completely independent of the hardware utilized in a specific network implementation. The remaining three levels are the only ones that need to be modified when changing network hardware.

<u>Network Layer</u>: Returning to our analogy, Prof. A's secretary must now interact with the telephone network "hardware" by either dialing Dr. B's telephone number, or by giving it to an operator. It is then the telephone company's responsibility to decide the proper routing of the call, make contact with the number called (by ringing the bell), or inform the caller that the called party is unavailable (by giving a busy signal). The analogous functions must also be performed by the network-dependent layer in a computer network, to provide message addressing and link establishment.

<u>Link Layer</u>: Once Prof. A's call has been placed, the telephone company also has the responsibility of deciding how to "link" her voice signal to Dr. B's telephone -- analog or digital, time or frequency multiplexed (or neither), transmission bandwidth and frequency, etc. The link layer in a computer network has the responsibility of deciding how a word of data is to be transmitted -- parallel or serial, synchronous or asynchronous, positive or negative logic, character- or packet- oriented protocol, baud rate (transmission speed), etc.

<u>Physical Layer</u>: The remaining layer essential to the completion of the exchange of ideas between Prof. A and Dr. B is the physical collection of wires (optical fibers, satellites); the electrical voltage levels (light pulses, microwave beams); and the mechanical handsets, junction panels, and switching stations necessary to form the physical connection between the telephones of Professor A and Doctor B. Computers often also employ the physical resources of the telephone system to complete the physical layer of their networks. However, many local area

networks utilize substantially different hardware to implement their physical link.

Of these seven layers, standards have already been adopted or proposed for only the first two or possibly three of them[7]. Standards for layer 1 include the common serial RS-232-C standard, the newer high speed serial RS-422/423 standard, and the CCITT X.21 protocol. Layer 2 protocols include HDLC and SLDC for serial links, CCITT X.25 protocol for packet switching, and the IEEE-488 interface protocols for both layers 1 and 2. X.25 is also being utilized at layer 3 for network control in some packet networks although it really was not intended as a layer 3 standard. It is hoped that by the end of 1981, standards will have been proposed for the remaining layers of the network heirarchy.

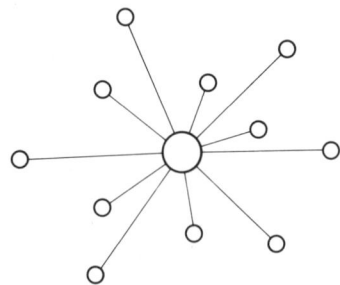

FIGURE 1.2. A star network topology.

In addition to defining the seven layers of computer network architecture it is useful to look briefly at some representative computer network geometries or topologies[9]. Four common topologies for large networks are the star, tree, mesh and ring.

The star topology, shown in Figure 1.2 is configured with a number of nodes (computers) attached to one central node. The central node is generally much more powerful than the peripheral nodes and most network traffic is assumed to occur between a peripheral and the central node. Peripheral-to-peripheral communications must be buffered through the central node in this configuration, often placing a heavy network burden on this computer. The star network is expensive to implement in terms of the physical layer, since each peripheral node must be connected directly to the central node.

Figure 1.3 shows a tree topology applied to the same node distribution as in Figure 1.2. The tree is the most efficient in terms of the physical layer, but it demands sophisticated

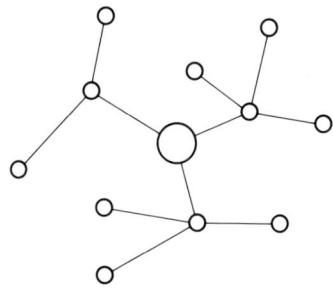

FIGURE 1.3.   A tree topology with the same nodal configuration as in Figure 1.2.

software algorithms and intelligent nodes to determine proper message routing.

A third topology, the ring (Figure 1.4), is not quite physically as efficient as the tree structure, but it is more flexible than either the star or tree topologies in that it requires no central node and no complicated data routing schemes. The data is simply circulated around the ring until it is received by the appropriate node.

The mesh (also called distributed) topology is illustrated in Figure 1.5. This mesh can be partially or fully interconnected, according to the traffic and reliability demands of a given network, and has the advantage over previously mentioned topologies of allowing the messages to be routed around a node if that node fails. There is, however, a price to be paid in terms of greatly increased hardware interface costs and complexity. In

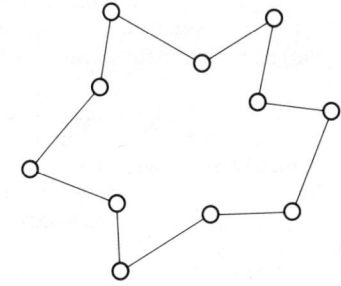

FIGURE 1.4.   A ring topology.

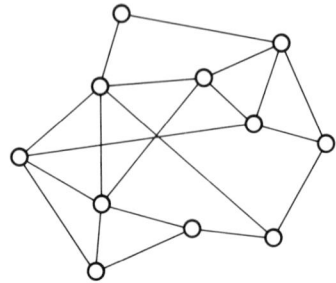

FIGURE 1.5.   A mesh topology.

spite of this disadvantage, mesh topologies are common in large commercial networks due to their higher reliability.

A disadvantage of all but the star topology is that the nodes in these networks are required to be active; they must act on every message they receive, either replying (if it is addressed to that node) or passing the message on. This can generate a great deal of network overhead and requires substantial network intelligence - a restriction, particularly if employed with nodes of limited computational capability.

A fifth topology, one that avoids the requirement for active nodes, is a "broadcast" topology. In such a scheme, each node is connected to a medium that is itself connected to every other node. Every message is then "broadcast" to every node simultaneously, and only those nodes to which the message is addressed need to respond in any way. This topology has the distinct advantage of isolating every node; if a node fails, it is the only node affected. Further, the physical layer can be optimized since each node need only be connected to the network medium and not directly to any other node. Examples of this type of implementation include the ALOHA satellite radio network in Hawaii, Ethernet™, and (to the extent that it can be considered a network) the General Purpose Interface Bus (GPIB or IEEE-488).

<u>Long</u> <u>Haul</u> <u>Networks.</u>   Distributed processing can be discussed in three general categories. The (chronologically) first, multiprocessing, was mentioned earlier in the section on mainframe computers. It is characterized by processing units that are separated by very short distances, operate at essentially cpu speeds, and are generally connected by a large number of data, address and control lines in a computer bus. The next form of distributed processing to make its appearance covered the other extreme in distance between processors and

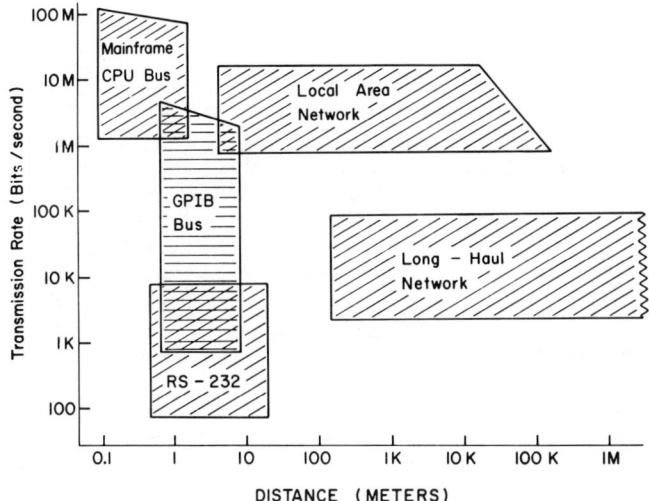

FIGURE 1.6. A comparison of the speed and distance limitations of various computer communications techniques.

speed of operation. It often spanned continents, operated at very slow speeds compared to cpu execution times. This form of distributed processing came to be known as long distance or long haul computer networking. The relative newcomer to distributed processing, at least in terms of widespread application, is the local area network. This form of networking, as its name implies, is optimized for path lengths of several meters to several kilometers, and the relatively short distances allow speeds approaching that of the cpu. The ranges covered by these three distributed processing approaches and by two other previously mentioned computer communications techniques are summarized in Figure 1.6[10].

Long haul networks first appeared in the mid 1960s. Concurrent with their early development, a report was published describing a military communications network in which messages were broken into short digital "packets" and broadcast in high speed serial bursts.[11] This technique was advantageous from a military standpoint since it offered high reliability because packets were easily retransmitted, and high security because messages were meaningless unless all of the packets associated with a given message were intercepted and properly reassembled. It was soon recognized that the packet approach had implications for computer networks, since packets from many messages could be interleaved on the same transmission medium, resulting in more efficient use of the communications circuitry; levelling of the "bursty" nature of computer communications; and fitting in well at the time with popular timesharing schemes for mainframe usage. The U.S. Department of Defense Advanced Research Projects Agency introduced ARPANET, the world's first computer network to employ "packet switching" techniques, in 1969[12].

Packet switching proved to be very workable in long haul computer networks and was almost universally adopted by many nationally sponsored computer networks throughout the world in the following decade. In summary, packet switching works as follows: A message destined for the network is disassembled into serial digital packets of fixed length. A network source and destination address is appended to this packet, along with a packet number, network control information, and error detection information. The packet waits for the next unused "slot" to appear on the network communications medium, and is "switched" into that slot. The network itself then controls the routing of the packet, often through a mesh topology, until it reaches its destination. If the packet was received with no error, an acknowledgment packet is sent; the packet is stripped of its addresses and control information; and it is reassembled into a message fragment by the receiving computer. If the packet was received with errors, a negative acknowledgment packet is sent and that message packet is retransmitted. Since no two packets are required to follow the same path through the network, this scheme is quite insensitive to the failure of an individual node, even in the middle of a message transmission: packets are simply transmitted around that node if at all possible. For a more comprehensive discussion of packet switching, the reader is referred to other sources[13].

As should be apparent from the above description, each node in a packet switching network must be rather sophisticated in order to assemble and disassemble packets, interpret addresses, and make network routing decisions. Often mainframe computers in long haul networks employ medium sized minicomputers as front end network processors to perform these functions.

One way to avoid complicated routing decisions is the use of a packet broadcast topology. ALOHA[14], the first such packet broadcast network, was pioneered in the Hawaiian Islands around 1971. In ALOHA, packets are bounced from a radio satellite to all network nodes at once. Only the addressed nodes must respond to a packet.

Packet broadcasting avoids routing complications, but it introduces "collision" problems. If two nodes begin transmitting a message at approximately the same time, their packets overlap and are said to be in collision with each other. Collision problems can be avoided by time division multiplexing schemes in which each node is allocated a specific slice of time during which only it can transmit a packet. This solution has been shown to be highly inefficient in networks where traffic loads fluctuate randomly. A more viable approach to collision avoidance and recovery in such a situation is found in network contention schemes. In a contention arrangement, all of the nodes are said to be "in contention" for the right to transmit. A given node waits to transmit until it observes an empty slot (time period) on the network. If another node begins transmitting in the same time slot, the two are in collision, and both nodes abort. Each then waits a short but random period of time before checking for another empty slot. The randomness of the time delay greatly reduces the possibility of a second collision between the same two nodes.

In the period since 1974, a number of mainframe and minicomputer vendors have begun the introduction of network hardware and software of their own[7]. In many cases, these networks employ packet switching techniques and utilize leased or dial-up telephone lines offering maximum data rates of roughly 50 K baud (50,000 bits/second), much slower than typical computer data rates. One of the major difficulties with these architectures to date is a total lack of standardization past the second or third of the previously mentioned seven layers of networking. However, as the use of computers continues its rapid increase, the need for a standard long haul network interface will make itself more strongly felt. This will result, with luck, in the production of an international networking standard for all seven network levels sometime in the next few years.

Local Area Networks. The proliferation of mini- and microcomputers in the late 60s and 70s made it quite common to find many small computers in a given laboratory or building. This often led to extensive duplication of costly peripheral devices such as mass storage media, hard copy devices, paper tape readers and punches, and the like. Each of these computers had to be complete stand-alone systems since there was often little compatibility between them, and no means for sharing information and resources among them.

By the early 70s, the need for some kind of "local area" networking scheme was obvious. Packet switching was well underway in long haul networks at the time, but this technique was considered inappropriate for local area networks because the required hardware and software was far too complicated and expensive to justify its use in connecting low-cost minicomputers over relatively short distances. A local area network at that time required simple and inexpensive hardware/software at each node to be compatible with the much simpler computer systems involved in a local area network. Simple network protocols were essential to keep from placing an undue burden on already limited computational capabilities with added network overhead, and high speed message interchange was necessary to handle the large amounts of data that could be generated in a real-time laboratory environment and still give the computer time to do something other than exchanging messages on the network.

The first steps toward satisfying some of the restrictions placed on local area networks were made by Hewlett-Packard when the company introduced its General Purpose Interface Bus (GPIB) in December of 1971. The GPIB, originally intended as a link for semi-intelligent instrumentation, is in many ways more like an extended computer bus than a computer network in that it has multiple data and control lines, and was designed for a maximum length of 20 meters. The topology of the GPIB resembles a star in the sense that a controller or central node arbitrates all network traffic. It differs from a star and resembles a multi-drop or broadcast network in that each node is connected to and can converse with every other node, but only addressed nodes (in this case, addressed by the controller as either "talker" or "listener") are involved in data exchange at any given time, eliminating the contention problems of typical broadcast networks. The GPIB did provide a solution to some of the unique problems of local area applications. Its parallel data lines offered bit-parallel, byte-serial data transmission at rates often approaching 200 K bytes/second, 400 times faster than common long haul data rates. The GPIB interface is simple, originally requiring a few dozen integrated circuits to implement, and by 1979 available as one or two LSI circuits. The major limitations in the GPIB are its limited primary address space allowing only 15 nodes, and the somewhat severe distance limitation of 20 meters. However, the availibility of LSI implementations of the GPIB interface can make it worthwhile for individual designers to work around the distance restrictions (i.e., see Chapter 5).

The first commercial attempt at providing true network capabilities for minicomputers was made in 1974 by DEC, the same company that introduced the first minicomputer nine years earlier. DECnet, as it was called, was more an attempt to provide long haul network capabilities for minicomputers than a solution to the unique problems of local area networks. DECnet was plagued with problems in its early stages, causing DEC to revise and re-release it as DECnet II in 1978. DEC is currently marketing DECnet III, a third generation network system for its family of minicomputers. Chapter 8 describes an implementation of DECnet in the context of nuclear generating stations covering the slightly-larger-than-local area of northern Illinois.

The same LSI technology that made microprocessors a reality is currently being applied to computer networking. The result, particularly for micro- and mini-computers is a drastic reduction

of the burden placed on the host computer by the requirements of
packet switching. Thus, the same techniques that were pioneered
on long haul networks can now be adapted to the requirements of
local area networks. Recognizing that this trend was inevitable,
researchers in the last five years have developed several packet
switching approaches for local area networks. Two of the more
promising of these approaches are discussed below.

In 1976, researchers at Cambridge University in Britain
implemented a local area network based on a ring topology. The
Cambridge Ring, as it was called, required roughly 80 integrated
circuits for its original hardware implementation. This network
has gone through several revisions until, in its current
configuration, it is availible as two LSI circuits[15]. Data
exchange occurs over twisted-pair wire at serial speeds as high
as 10 MHz, and nodes on the ring can be separated by as much as
100 meters. The ring functions by circulating a series of empty
message packets around its perimeter. A node sends a message by
placing it in one of these packets and marking that packet as
full. The 38-bit packets contain six control bits, two eight-bit
addresses for source and destination, and sixteen data bits.
After making one loop around the ring, and presumably being read
by the destination device, the packet is marked as empty by the
node that originally sent it. This prevents unread packets from
circulating indefinitely around the ring and keeps network
maintainance to a minimum. The Cambridge Ring nicely fulfills
the necessary requirements of speed, simplicity and cost for
local area network interfacing. One major drawback, as mentioned
earlier, lies in the fact that a ring topology requires active
nodes, and a failure at any node serves to incapacitate the
entire ring.

An alternative to the Cambridge Ring is Ethernet[16], a
contribution from the Xerox Corporation. Like the aforementioned
Ring, Ethernet was introduced in 1976. Also a packet-oriented
scheme, Ethernet uses a broadcast topology and a network
contention scheme similar to the one described in the section on
long haul networks. Its name bears reference to the fabled
luminiferous ether that was at one time thought to permeate space
and support electromagnetic radiation. Like the earlier ether,
Ethernet's ether (a co-axial cable as employed in cable TV
systems) acts as a completely passive medium, in this case, for
the transmission of digital data. This implies that failure at
any node will affect only that node.

The original specifications for the Ethernet prototype
included a distance of 1 km, serial data rates of 3 MHz, and an
addressing maximum of 256 nodes. Current definitions call for a
packet length of 4000 bits. Ethernet has already been adopted by
the National Bureau of Standards for its internal use; Xerox,
Intel and DEC presently are working together to produce an LSI

version of the Ethernet interface and to push for the acceptance of Ethernet as the first recognized local area network standard[10]. Predictions abound that plug-compatible high speed local networks based on Ethernet or its equivalent will be available as early as 1983.

It is obvious that with computational capabilities and needs as diverse as they already are and will continue to be, no local area or long haul network alone will suffice to solve all computer intercommunications problems that will arise. In addition, it is inconceivable that one network type will ever become universal. It is desirable, therefore, to have a method of passing information between differing networks. This is provided for (conceptually, at any rate) with the idea of a gateway node. The only function of a gateway node is to appear to be a node for one network on one side, and a node for a second network on another side. With such gateways, one can visualize an entire network of local area networks, all connected through a common long haul network. Such devices make it entirely plausible to predict a future in which no computer is an island entire of itself (with apologies to John Donne[17]).

SUMMARY

In the thirty some years since ENIAC, electronic computation has developed into a force that has already begun to revolutionize society. In these few pages we have followed computers as they grew from massive but rather stupid calculational machines to the ever more powerful mainframe systems of the late 50s and early 60s. We have seen them shrink into minicomputers in the 60s and then into microcomputers in the 70s, all the while finding their way into diverse and unexpected corners of our lives (this chapter itself was composed and typed on a microcomputer based word-processor). If current trends are any indication, computers will continue to increase in power while they decrease in size. But even as the 60s can be considered the decade of the mini and the 70s the decade of the micro, the 80s will not merely offer more of the same, but will become the decade of the network, in which computers join forces to make their information more readily available when and where it is needed.

The remaining chapters in this book detail some first steps in the use of computer networks to aid in doing chemistry. Keep in mind while reading these chapters that many of the ideas represented here were formulated and put into practice by chemists at a time when even computer scientists were still hotly debating the proper structures for computer networks. Not all of the vocabulary in these chapters will be used in exactly the way it has been defined in this chapter. Not all of the vocabulary will even have _been_ defined in this chapter. This is not surprising, but welcome, because an area as dynamic as networking is today should be represented by a dynamic vocabulary where the words change to fit the ideas they represent, at least until the ideas themselves have crystallized.

REFERENCES

1. Much of the factual material for this discussion was gleaned from an excellent review of the last 50 years of electronics development published as: *Electronics,* 53, (1980).

2. ibid., 185.

3. ibid., 381.

4. M. Marshall, *Electronics,* 53, 39 (1980).

5. J. Matisoo, *Scientific American,* May, 50 (1980).

6. G. A. Kildall, *Dr. Dobbs' Journal of Computer Calisthenics and Orthodontia,* 5, 6 (1980).

7. A. Durniak, *Electronics,* 51, 107 (1980).

8. G. L. Leger, *Electronics,* 52, 89 (1980).

9. D. W. Davies, et al., *Computer Networks and Their Protocols,* Wiley-Interscience, New York, 1979 Chapter 1.

10. R. W. Comerford, *Electronics,* 53, 40 (1980).

11. P. Baran, *IEEE Trans. on Communications Systems,* CS-12, 1 (1964).

12. F. E. Heart, et al., AFIPS Conf. Proc. Spring JCC, 36, 551 (1970).

13. One exellent source on packet switching is reference 9, chapters 2,5,6,7.

14. R. Binder, et al., Proc. Nat. Computer Conference, 203 (1975).

15. K. Smith, *Electronics,* 53, 80 (1980).

16. R. M. Metcalf and D. R. Boggs, *Comm. of the Assoc. of Comp. Mach.,* 19, 395 (1976).

17. J. Donne, Devotions Upon Emergent Occasions, (1624) as cited in: E. Simpson, *John Donne Selected Prose,* Oxford University Press, London, 1967.

2

A GENERAL PURPOSE MICROCOMPUTER FOR LABORATORY AUTOMATION IN A HIERARCHIAL ENVIRONMENT

W. Stephen Woodward, Charles N. Reilley

INTRODUCTION

    The microcomputer-based instrumentation automation system described here is an outcome of a continuing development effort in our laboratory to apply microprocessor technology to analytical instrumentation. Since its inception late in 1972, this effort has been directed at the investigation and design of microcomputer systems well suited to the general nature and requirements of contemporary and novel laboratory instrumentation and its researcher/operator. The salient design features include extreme low cost, modularity, optimized I/O structure for real time experimental control and data acquisition, economical large random access memory implementation, flexible video display generation for operator interaction and data presentation, hardware program bootstrap, compatibility with a host computer support system, as well as provisions for local mass storage and hard copy generation. Application to numerous specific instruments have been made and it has also been used as a stand-alone instructional unit[1].

THE HIERARCHIAL SCHEME

    The general nature of the computer scheme to be implemented at UNC is illustrated in Figure 2.1. The role of the microcomputer is to provide a means for direct interfacing to a variety of laboratory instruments and hence to remove the need for the host computer to provide real time functions of instrumental control and data acquisition. Our philosophy in this whole

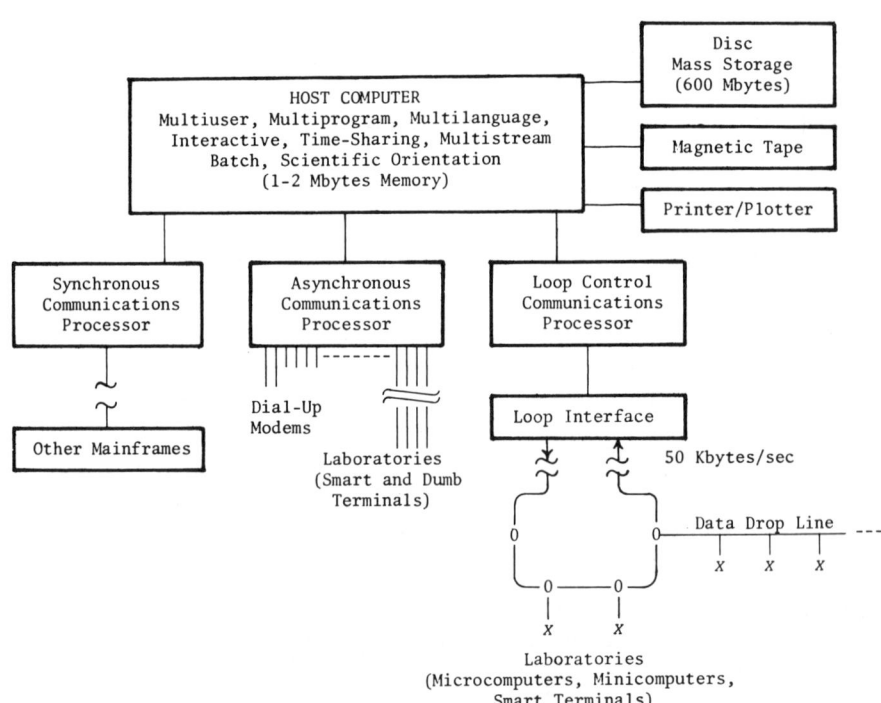

FIGURE 2.1.   The hierarchial scheme illustrating the placement of the general purpose microcomputers (X).

scheme is to have as little load placed on the host computer as possible. Thus the microcomputer stations can provide a variety of functions in addition to data acquisition and control, including local data storage, graphics (hard and soft), printing, data work-ups, etc. Occasions of course do arise where the power of a host computer, because of its peripherals, large storage, or inherent number-crunching power, is needed. In addition to their laboratory functions, the microcomputers also can act as smart terminals to the host computer, providing local storage, graphics, etc.

In our environment the host computer will serve the needs of a variety of chemists, including the theoreticians. Particularly attractive for this purpose are the 32 bit computers with enhanced scientific computing power. Among these are the virtual memory machines (Data General MV8000, Digital Equipment VAX 11-780, IBM 4331-II or 4341, or Prime 750) or the direct addressing machines (Perkin Elmer 3240 or SEL 32/7780). Future considerations are the possible addition of array or vector processors as well as high-resolution graphics. For improved performance communication processors are present to off-load the host processor and hence provide the capability for more active on-line terminal users without bringing the host computer to its knees. Note that most of the laboratory microcomputers will communicate to the host via a quite efficient and fast loop-oriented local data network. This scheme utilizes a small coaxial cable which loops from laboratory to laboratory and contains a local T-tap at each laboratory, to which is attached a data drop-line of 100-200 feet which in turn connects several (up to 20) microcomputers in a daisy chain fashion. The coaxial loop is quite tolerant of long lengths, can contain 50-60 T-taps, with the result that the whole loop communication scheme can support numerous terminals (i.e. 256).

DESIGN FEATURES OF THE GENERAL PURPOSE MICROCOMPUTER

Some of the salient features of the resulting design are as follows:

1. <u>Extreme low cost</u>. If the computer is to become truly a commonplace measurement tool, its cost to the experimenter must be in line with existing general-purpose laboratory aids. A target figure of $1200-$1600 component cost for a basic but complete system for high-performance data acquisition and interactive analysis has been achieved through microcomputer technology.

2. <u>Modularity</u>. A viable computer-assisted chemical analysis

data acquisition environment is often characterized by a quite rapid rate of application turnover. Hence, a general-purpose data acquisition system is likely to be confronted by a succession of applications, each with differing requirements. A system configuration suitable for this environment must either be capable of handling most expected application demands in some standardized configuration, or else must lend itself with great ease to a "tailoring" process so that it may be "scaled" either up or down as the occasion demands. In an effort to maximize the potential efficiency of capital investment, the system described here exploits the latter method to an advanced degree.

Modular organization at the system-function level has been adopted to facilitate assembly of application-specific configurations. In this approach, high-level system components such as the central processing and system controller unit (CPU), large (64K byte) memory arrays, and peripheral input/output controllers are each implemented as single 4.5 x 10 inch etched-circuit cards. These cards are interconnected via insertion into a 44-conductor, crosstalk resistant parallel backplane bus, (which can accommodate up to 22 such units), to form the basis of complete systems.

3. Optimization of input/output structure for experiment control. Careful examination of real-time applications in the chemical laboratory indicates that a small computer dedicated to instrumentation will usually run out of response time and data acquisition speed capability long before its computational capacity is exceeded. The computation associated with experiment control is usually associated with post-experiment data interpretation and need be completed only in amounts of time commensurate with the patience of the operator. The speed of actual data acquisition, however, is usually dictated by the dynamics of the system under study. Hence, if the computer system is too slow in that regard, the work simply cannot be performed.

For these reasons, the design of a general-purpose data acquisition system must achieve a highly versatile I/O structure possessing speed and generality for real-time operations disproportionate to the computational power of the CPU. Instrumental control and data acquisition related input/output has therefore received heavy emphasis in this system design. This emphasis includes selection of an overall system design which includes features useful in the provision of such functions [e.g. easily used direct memory access (DMA)], the design of a large variety of standard "measurement-oriented" I/O modules, (e.g. high performance analog-to-digital and digital-to-analog conversion and timing modules), and the design of a system format supportive of implementation of specialized instrumentation interfaces, (e.g.

provision of well-regulated standard supply voltages, and adoption of a popular circuit-card format insuring commercial availability of prototyping hardware).

4. <u>Economical, expandable implementation of large random access memory arrays</u>. Analysis of various categories of measurement-automation applications reveals frequent requirements for "memory-intensive" system designs - either due to large control programs or the generation of extensive data arrays. 16K dynamic memory chips were therefore chosen in the UNC-Chem design as the basis for main system storage because of the superior cost, space, and power requirements of these units (less than 0.03 cent, 1.5×10^{-5} square inches, and 20 microwatts per bit, respectively), when compared to any currently readily available competitive technology.

Memory system utilizing dynamic chips must, however, incorporate some techniques for providing the periodic refresh activity needed to maintain the integrity of stored data. The methods used to implement memory refresh in most micro-computers using the dynamic technology may be characterized as <u>asynchronous</u> in the following sense: In generally used refresh techniques, necessary refresh cycles are "stolen" from normal memory activity upon demand from a free-running timing circuit. These stolen cycles occur independently of other system activities and are, thus, asynchronous to those activities. In the general run of micro-computer applications, this method of memory maintenance is simple to implement and efficient in terms of average memory availability. A difficulty arises, however, in the context of high-performance instrumentation control from the following circumstance: If, in the course of system activity, a memory access request (as might be generated in an attempt of the CPU to fetch an instruction) should occur simultaneous to a refresh cycle "theft", asynchronous refresh logic would defer granting of the system access request until the refresh cycle was completed, thus anomalously increasing (usually doubling) the time needed to satisfy the deferred request. Because such conflicts occur unpredictably during the progress of experiment control, unavoidable timing indeterminacy is introduced into computer/instrument transactions. While the resulting indeterminacy is small, and thus important only for relatively high-speed experiment control, the performance we desired for the UNC instrument made it intolerable.

This effect is avoided in the UNC micro-computer system by two measures. First, timing for refresh is derived directly from the main system timing reference and is thus <u>synchronous</u> with all other system activity. Secondly: refresh is performed at an <u>average</u> rate far higher than specified by the memory-chip manufacturer as needed for the maintenance of data integrity. This fact allows memory access for refresh to be given a priority <u>lower</u> than that given CPU and I/O-originated requests. Thus, in the event of an access conflict, it is the <u>refresh</u> request that is deferred guaranteeing that the experiment-related transaction will <u>always proceed on schedule</u>. The very high average rate of refresh insures that, even if a large number of such refresh-cycle denials occur, data integrity will not be threatened.

    5. <u>Flexible video display generation</u> is incorporated as an integral function of the CPU module. Effective interaction of the researcher with computer-controlled instrumentation is dependent upon timely and meaningful presentation of acquired data and experiment status.

    One of the most consistently successful means of data presentation is the CRT graphic display. Early in the design of this microcomputer system, provision was made for an economical graphic display with useful resolution using low cost video monitors (or even lower cost consumer-grade TV's) as the display medium. Display format is a 256 x 256 point map, permitting presentation of both graphic (e.g. spectra) and alphameric (e.g. system messages and numeric tables) data. Central to the feasibility of this display design approach, (due to the need for a 65,536 bit memory area for service as an image buffer) is a method for efficient implementation of large memory arrays. Use of dynamic 16K MOS memory components yields a display-related memory cost of $30 at current prices. That figure, when combined with the cost of a suitable monitor and other display hardware, leads to a total cost for this versatile operator interaction resource of under $150. At this level of expense, mixed graphic and alphameric display capability becomes economic as a universal basic peripheral of low cost systems. An interesting feature of display generation is that memory references needed for display maintenance are simply those occurring in the course of memory refresh. The devices used in system memory are refreshed by the execution of read cycles. Refres<u>h</u> maintenance, therefore, is performed very well simply by the previously described memory activity associated with display generation. Thus, the television is driven largely by logic which would be needed in any case merely to preserve memory contents. The only additional circuitry needed consists of an eight-bit shift register to serialize the video output, and provision for T.V. synchronization-pulse generation.

    6. <u>Compatibility with a host support system</u>. Because the UNC-Chem micro-computer was configured from the outset to be well suited for duty as a component in a multi-laboratory system,

features have been incorporated in its design to facilitate such operation. "Bootstrapping" of the terminal computer into operation from a host system (i.e., initial loading of controlling programs), transfer of data acquisition control programs from the host library, data transfer, and other facets of computer-computer communication are easily and efficiently performed.

7. Hardware-implemented program bootstrap. Because of the envisaged rapid turnover of application programming, all of a micro-system memory is composed of read-write storage, rather than partly read-write and partly read only. This organization provides maximum flexibility in program implementation and change. However, an obvious requirement is generated for a means of initial program entry in the event of memory loss. The availability of the program-independent DMA input-output, combined with a small amount of additional circuitry, permits the UNC micro-computer prototype to accept initial program input from a variety of sources including cartridge or cassette mag-tape, flexible disk, UART, or a supporting host computer. Specific "bootstrap device" selection is accomplished by a jumper wire connected from the CPU card to the appropriate device controller. The bootstrap or FILL mode is initiated automatically upon power up, manually, via a front-panel switch, or under program control. The FILL mode is terminated and program execution initiated by end-of-record status within the bootstrap device controller.

HARDWARE COMPONENTS

System elements and modules of the UNC-Chem microcomputer system group naturally into a number of functional categories, although a single component will occasionally offer utility in, and therefore logically belong to, more than one grouping. A listing organized in this fashion of available system elements including approximate cost, construction effort (in man days), and functional description, follows.

Central Processing Components.

1. System Housing ($180, 2 m-d): Built entirely from commercially available modular components, the system housing includes power supplies, card guides, connectors, and backplane bus.

2. CPU Card ($175, 1.5 m-d): Carries 8080A microprocessor

chip and performs functions associated with microprocessor support, bus management, Direct Memory Access, Video Display Generation, memory control and initial program load (automatic bootstrap).

   3. 64K-Byte RAM Card ($300, 1.0 m-d): Carries 32 16K by 1 bit dynamic Random Access Memory devices (Intel 2117 or similar) to provide 65,536 bytes of power-conservative program and data storage. Cycle time is 750 nsec.

Real Time Data Acquisition and Experiment Control.

   1. Successive Approximation ADC Card ($130, $170, or $270, 0.75 m-d): Accepts any one of a series of pin-compatible 12-bit ADC modules of differing maximum speeds (DATEL ADC-H series) to support 12-bit throughputs of 45K to 330K conversions/sec and 8-bit throughputs of 57K to 430K. Control signals are provided for up to 16 channels of parallel track/hold and for interface to the high-speed averager card. Provision is made for use of this ADC card with the I/O sequencer module.

   2. Quad Track/Hold Card ($135, 0.75 m-d): Implements 4 independent track and hold amplifiers with associated analog multiplexing so that, at the onset of a conversion scan cycle by the ADC card, all four input signals are simultaneously captured and sequentially presented to the ADC for conversion. Up to 4 such cards, for a total channel count of 16, may be connected to one ADC. Specifications include acquisition time of 3 μsec to 0.01%, aperture uncertainty of 20 NS, and multiplexor delay of less than 1 μsec to 0.01%.

   3. Hardware Averager Card ($40, 0.25 m-d): This card implements a direct interface to the ADC by means of which 8-bit conversions may be added to (or subtracted from) multiple-byte-precision memory accumulations. In this way, enhancement of experimental-data signal/noise ratios may be achieved through ensemble averaging. 8, 16, 24, or 32 bit sums may be computed with number of summation bits, as well as selection of addition or subtraction, under program control. Time required for summation update is 3 μsec/byte.

   4. High Speed I/O Sequencer Module ($55, 0.5 m-d): This card provides a versatile means of generating time-related sequences of computer-experiment interactions through direct memory access at speeds far exceeding the capabilities of the

microprocessor to manage such activity. The I/O sequencer utilizes a 16-bit variable modulus counter driven from the 2 MHz system reference to control timing between experimental "events" with an accuracy of 0.01% and a resolution of 0.5 μsec.

Upon expiration of an inter-event period (as determined by overflow of the counter or by receipt of an external trigger), initiates the generation of a sequence of up to 8 I/O control pulses. Pulses appear on individual points available on the sequential programming module and are routed to those elements in the controlled instrumentation system which are to participate in the I/O program.

For any given event a subset of the 8 pulse lines are selected for actuation by an 8-bit program mask byte loaded into the program shift register from microcomputer memory during the interval between events. In addition to sequencer operation, this module can also be used as a 16- or 32-bit event timer.

5. <u>Quad 12-Bit DAC ($165, 0.5 m-d)</u>: Four 12-bit, independently latched analog output channels are provided by this card. Approximately 5 μsec are required for the re-load and settle of any channel (under DMA). The transfer of new 12-bit values to the DAC buffer register are performed as parallel "jam" transfers so that output "glitches" are minimized. Operation under I/O sequencer control is possible. The inclusion of a plotter-pen control bit on this card makes it useful as an XY plotter interface. Select address strapping permits up to 4 of these quad DAC modules to co-exist in a single system.

6. <u>Integrating ADC and 20 Hz Clock ($60, 0.5 m-d)</u>: This peripheral implements three channels of 16-bit resolution, auto-zero-compensated analog to digital conversion with a maximum aggregate throughput of 20 points/second. The span of each channel may be selected (via input resistor network) independently of the other two. Inputs are fully differential (60 db common-mode rejection) with a 100K ohm/volt impedance and high "survivability" (i.e., application of 110 VAC to inputs without damage). Normal-mode rejection of 60 Hz-related noise is greater than 66 db (limited by power-utility frequency accuracy).

<u>Operator Interaction And Data Presentation Peripherals</u>

1. <u>Video Display (TV) ($100, 0.25 m-d)</u>: The major medium available for immediate communication to the operator of system and experiment status and data is a 12-inch video display implemented <u>via</u> a consumer-grade solid state television receiver modified to accept EIA-standard composite video (or, equivalent but

more costly, a standard CCTV monitor). The output medium so implemented is both a graphic display of 256 x 256 resolution and an alphameric display with a format of 42 lines of 64 characters each (2688 characters, total). Graphics and alphameric text may be freely mixed. Display "polarity" (i.e., white-on-black or black-on-white) is program selectable.

2. <u>ASCII Keyboard ($160, 0.06 m-d) and Controller ($25, 0.2 m-d)</u>: A parallel-interface ASCII keyboard (Cherry Inc. #B70-61AA) with companion controller module, this peripheral provides a flexible means for input of operator commands to the microcomputer. The controller will support a variety of parallel keyboard interface requirements.

3. <u>Teleprinter ($800-$1800) and Interface ($25, 0.25 m-d)</u>: The wide variety of available teleprinter terminals affords the basis for flexible operator communications where hard copy is desired. The interface implements a general-purpose asynchronous serial link (UART-based) and accepts 20-m.a. current loop or RS232-standard signals of 110, 300, 600, 1200, 2400, 4800, or 9600 baud. All common character formats are accomodated.

4. <u>XY Recorder ($1500, 0.5 m-d) and Controller ($105, 0.5 m-d)</u>: Software has been developed which permits the use of standard analog XY recorders (Hewlett Packard Inc. #7040) as versatile generators of annotated graphical hard copy.

Magnetic Media Storage

1. <u>8-Inch Density Flexible Disk ($650, 1.5 m-d) and Controller ($75, .75 m-d)</u>: Utilizing 1 to 4 of the many industry-standard 8" drives (Shugart Associates 801/851, Siemens FD120/240, MFE 501/701, etc.), a reliable, high-speed, random-access, name-oriented memory is provided with an on-line capacity ranging from 650 Kbyte (one single-sided drive) to 5.2 Mbyte (four double-sided drives). Average transfer rate is 50 Kbyte/sec. for arbitrarily large (<64K) records. Worst-case random access is dependent upon head-seek speeds of the particular drive used but seldom exceeds one second for directory-driven retrievals. Recorded data is encoded in a run-length-limited code similar to GCR (Group Code Recording). Such codes display greater decode-timing margins than the more popular MFM (Modified Frequency Modulation) and related techniques. Greater resistance to bit shift and jitter on readback and consequently superior data reliability result.

2. 1/4" Data Cartridge Drive ($700, 1.5 m-d) and Controller
($70, 1 m-d): This peripheral provides a high-performance
medium for rapid storage and retrieval of especially large data
volumes. It consists of the 3M DC-300 data cartridge used with
a Mohawk Data Sciences 2021 transport and in-house read/write,
motion control, and formatting electronics. The recording
medium is formatted as a directory-driven random access storage
area within which as many as several hundred files may be
accessed by user-assigned alphameric names of arbitrary length.
Files may consist of any number of variable length records up to
a maximum of 700K bytes per file. Total cartridge capacity is
2.8 mega-bytes organized as 4 independent tracks. Provision is
made on the controller card for the management of up to 4
transports, yielding the capability for an on-line capacity of
up to 11.2 Mbyte. Recording density is 1600 bits/inch and data
transfer occurs at 6K bytes per second. During recording, a
read-while-write check is performed of data record integrity.
Worst-case random-access delays are proportional to the fraction
of cartridge storage capacity utilized and are approximately 30
millisec. per Kbyte. Thus, access times over a 200K byte system
would not exceed 6 sec.

3. Audio Cassette Recorder ($35) and Controller ($40, 0.5
m-d): Based upon a simple FSK (1200/2400 Hz) extremely speed-
tolerant (± 50%), synchronous data encoding format, this peripheral allows the reliable storage and retrieval of substantial
data/program files (up to 1.4 megabyte per C120 cassette - both
sides unformatted) at respectable speed (200 bytes/sec.) using
entirely unmodified consumer-grade audio cassette recorders.
Sequential access delay is 3 seconds maximum (recorder start/
stop time). Random access by file name ranges from tens of
seconds to tens of minutes depending upon cassette utilization
and the amount of manual intervention supplied by the operator
in the form of fast forward tape movement, etc.

Communication with Shared-Resource Computing Hardware

1. Parallel Digital Communication (Lab-Box) Interface ($40,
0.5 m-d). Much of the development of UNC-Chem micro-computer
software has been performed using available Raytheon minicomputers in the role of cross-assembly and linking down-load processors. The ability to make use of convenient source-handling
peripherals (card reader, line printer) has proven to be a great
asset. In addition, the use of minicomputer-maintained magnetic
disk memories as program libraries and mass data storage has
been invaluable (particularly before the availability of magnetic

media priced compatibly with the microcomputer) as has the
ability to access shared minicomputers as computational resources

The Lab-box interface implements the micro-computer side of
a digital communication link providing a 10K byte/sec, automatic
error correction via retransmission, down-load, and cold-start
(bootstrap) capabilities.

2. Asynchronous Serial Interface ($25, 0.25 m-d): An
alternate means of computer/computer communication is provided by
this module. 20 ma. current loop, RS232 and TTL compatible, the
serial interface supports full-duplex communication at rates up
to 9600 baud via direct wire or modem link.

3. Stream-Multiplexed, Packet-Oriented Serial Link ($50, 0.5
m-d): This module implements the microcomputer side of a ring-
organized, integrated communication system designed for a hier-
archical multilaboratory computer network with the following
properties:
Handles all system communication via one party-line cable;
Bit-serial;
Unidirectional data flow (circular bus topology);
Data and control are transmitted in 10-bit "packets" at 0.5
mega-baud to achieve a net transfer rate of 50K byte/sec;
End-to-end error control is maintained in all bus activity;
Connection of terminals to bus is via signal-regenerating
"T-TAP"/repeaters - allowing virtually unlimited cable length
and number of terminals;
Ground isolation (opto-isolators in taps);
Selection of terminals by host-computer polling;
Polling transaction is one 20 micro-second packet.

SOFTWARE DEVELOPMENT TECHNIQUES AND RESOURCES

Crucial to the utility of any computer system is, of course,
the ready availability of effective means for the development,
de-bug, and access of general-purpose and application-specific
software. Major categories of software support needed for the
full and efficient utilization of computer techniques are:
1. Machine-oriented-language software development
   (assemblers).
2. Macro-level utilities (floating-point and multiple-
   precision arithmetic, transcendental function packages,
   I/O handlers, I/O formatters, graphic-output generators,
   command-string parsers, etc.)
3. General system utilities (text handlers and editing

resources, relocatable and memory image libraries and loaders, linking load-module generators, general file maintenance, etc.)
4. Problem-oriented-language software development (compilers, interpreters).

Early in the organization of the UNC-Chem microcomputer effort, the decision was made to develop an interim software development resource around available minicomputers. The resulting package of cross-assembly relocatable library, and down-loader were economically implemented through maximum use of existing minicomputer utilities and served us well for several years.

More recently, however, a complete set of microcomputer-resident development utilities have been completed and comprise the items in Table 2.1.

TABLE 2.1   ASSEMBLY-LANGUAGE SOFTWARE DEVELOPMENT UTILITIES

I. SOURCE TEXT EDITOR

   A. Character, user-defined field, and line-oriented edit capabilities.

   B. Total independence from numeric edit parameters (e.g. line numbers and character counts) even for multiparameter commands (e.g. block move) through unique cursor-position push-down stack.

   C. Compatability with named-file mass storage of source.

   D. Compacted internal source representation.

II. ASSEMBLER

   A. Pass-and-a-half operation. Only one pass needed over source for both object text and listing generation but full freedom in forward-reverse references.

   B. Generates relocatable linkable code for convenient and efficient use of large utility libraries.

   C. Fast. Approximately 60 source lines per second.

   D. Mass storage manipulation.

III. LINKER/LOADER

   A. Automatically searches multiple large utility and user

TABLE 2.1 continued.

        libraries, allowing convenient access to these resources.

    B. Performs optimized-itinerary sequential library searches. Typical linking-load times of 20 seconds from large libraries.

    C. Effective use of name-file mass storage for both library access and absoluted-module output.

IV. RELOCATABLE LIBRARY AND LIBRARY MAINTENANCE UTILITIES

    A. Large (>60 module) library of general-utility modules including:
        1. Multiple byte fixed and floating point math.
        2. Most common math functions (trig, log, exponential, etc.)
        3. Formatters.
        4. I/O handlers - device independent calling sequences.
        5. Graphics generation.

    B. Library maintenance utilities for creation and maintenance of both global and user libraries.

    C. Extensive use of named-file mass storage.

V. DEBUG MONITOR

    A. Compatability with relocatable environment - acceptance of address expressions.

    B. Memory and register content inspection and modification.

    C. Transparent breakpoints.

    D. Single-step instruction execution.

At the date of this writing, approximately 17,000 source lines of such code have been generated providing a rich environment for the development of applications software.

Among the tools produced in this way is an instrumentation-oriented BASIC interpreter which possesses a unique combination of data-file and graphic image manipulation capabilities. The BASIC language has proven to be a useful resource for both the instructional environment and for researchers unfamiliar with the powerful but arcane methods of the assembly-level programmer.

APPLICATIONS

A number of applications have already been made of the UNC-Chem microcomputer, some rather simple, and some rather complicated and requiring high performance data acquisition and/or control. Early surveys have been published.[2,3] Simple applications can follow along lines already described[4] and include autotitrations with computer control of motor driven burettes and acquisition of data from a pH meter, data logging from IR and UV-Visible spectrometers, in which multicomponent analysis of samples is achieved via use of multiple wavelength data for standards and samples, interconnection to a high resolution UPS unit containing a multichannel analyzer where data is simply transferred from the analyzer to the microcomputer for subsequent data analysis and for X,Y plotting, etc. A number of chromatographic applications have been accomplished where auto-integration, peak detection, and quantitation via standards are standard routines. In HPLC multiple fixed wavelength detectors, scanning detectors (via stepper motor control), gradient solvent mixing, and a variety of autosamplers are features which have been brought under control of the microcomputer. Autosamplers and a wide variety of detectors have also been interfaced for gas chromatography applications. Other related applications include scanners for TLC plates and electrophoresis gels. A DuPont 650 ESCA spectrometer has been interfaced as well as a Phi 548 ESCA (X-ray and UPS sources) and Auger spectrometer. The software routines of ESCA were written in assembly language[1] and those for Auger accomplished by adding a single command to BASIC and allowing a user to program the rest in BASIC. The UNC-Chem microcomputer has also been applied to control transmitter pulses and acquire and ensemble average quadrature data in a FTNMR spectrometer[1] as well as a variety of electrochemical techniques (cyclic voltammetry, double potential step chronoamperometry, staircase voltammetry, and FT). Communication software has been written to permit the microcomputer to act as an intelligent terminal to host computers via dial-up modems, permitting data files created via BASIC to be converted and shipped to the host (i.e., IBM CALL-OS) for data crunching. Ten microcomputers have been assembled and used in an instructional course for both numerical analysis and for instrument interfacing.[1] Work is currently in progress to adapt the microcomputer for FT-ICR (where frequency synthesis is controlled and quadrature detection data is acquired), for fluorescence life-time measurements (2 MHz 8-bit transient digitizer on a plug-in board with rapid transfer to microcomputer memory and ensemble averaging), and for single photon or ion counting where a 12.5 nsec time window is used in connection with 32K memory permitting measurement of multiple arrival times at an effective acquisition rate of 80 MHz. Spectral search systems for ^{13}C-NMR, IR, and mass spectroscopy is nearly complete.

DOCUMENTATION

A three-volume set describing in detail the construction, software, and theory of operation of the UNC-Chem microcomputer system has been placed in the UNC Chemistry library and can be obtained via standard interlibrary loan procedures.

ACKNOWLEDGEMENT

A number of graduate students have contributed to development of the UNC-Chem microcomputer. Particular thanks are due to Stephen Brandt, Bennie Good, Judy Hinderliter-Smith, Jerry Koontz, T. H. Ridgway, David Smith, Alan Uthman, and Gene Woodard.

REFERENCES

1. W. S. Woodward and C. N. Reilley, *Pure and Appl. Chem.*, 50, 785 (1978).
2. W. S. Woodward, T. H. Ridgway, and C. N. Reilley, *Analyst*, 99, 838 (1974).
3. C. N. Reilley, W. S. Woodward, and T. H. Ridgway, "Micro-Computers - A Future Solution to Many Problems?", in *Information Chemistry*, S. Fugiwara and H. Mark, *Eds.*, University of Tokyo Press, 1975.
4. A. Carrick, *Computers and Instrumentation*, Heyden and Sons, Inc., Philadelphia, 1979.

3
FORTH, A FOURTH GENERATION LANGUAGE IMPLEMENTS A NETWORK

R. E. Dessy, David J. Hooley

BACKGROUND

The computer network in the Computer Aided Research Equipment (CARE) Group at Virginia Polytechnic Institute consists of 12 DEC PDP-11 systems. These are tied together by point-to-point 9600 Baud RS-232 transmission lines, with three host computers and nine satellite computers. Inter-host as well as host-satellite communication is supported. The host computers are 28K DEC LSI-11s, equipped with major peripherals such as disks, line printers, digital plotters and interactive graphics. The satellites are 16K DEC LSI-11 based computers. All are equipped with 16 channel multiplexed ADCs, dual DAC output, and parallel I/O ports. The terminals throughout the system are a mixture of CRTs and DECwriters. The hosts are used for program preparation, and the source of down-line loaded code. They also receive raw data, condensed reports, and control information from the satellites. Long term data storage and correlation efforts are performed on the hosts. Time-critical tasks are reserved for execution at the satellites, which are not burdened by excessive multitasking or the control of demanding operations such as graphics or digital plotting. All debugging takes place at the satellites. Each satellite assumes responsibility for from one to three instruments in its immediate environment. This spreads the cost of each satellite facility over several applications, yet maintains ease of programming and speed of response.

The net was originally implemented using a subset of Decnet, called Remote. This permitted operation under RT-11 (a fast, efficient but fragile operating system) of a multiuser program allowing a mixture of terminals and satellite computers. Use of background for compilation/assembly was limited to only one user at a time. The satellites ran a simulated version of the operating system (OS). The network supported Fortran and Assembler languages.

The strength of the approach was its ability to provide

higher level language support to multiple users in an environment that permitted easy installation and running of real-time tasks. RT-11 does permit the user to have a more intimate control of the system functions than he has with executives such as RSX-11M, which isolate the user from direct contact with system functions. The weakness of this approach lay in the lengthy and often convoluted procedures required to achieve a linkage between Fortran object modules to manipulate data, and Assembler object modules to acquire the original data and display the results. Difficulties were also encountered when the simulated OS in the satellites failed to exactly replicate the OS running on the host. As new revisions of the system became available, it was obvious that 28K was not enough memory to simultaneously support three terminals at the host, and four satellite computers at each host.

Overlapping the latter stages of network development under RT-11/Remote was the initial examination of several public domain versions of a programming system called Forth. This led to the acquisition of a license to run a commercial version of the language, and the re-implementation of the net to use it. Eventually several versions of the language were produced in our own laboratory targeted for PDP-11 computers, as well as others.

Forth is an applications oriented language ideally suited for use in satellite computers supporting multiple real-time tasks. It is the function of this chapter to focus on the structure of the language, and on its application to a network.

The advantages of Forth lie in its block oriented structured approach to applications software and its inherent small memory size and high execution speed.

Under RT-11 (V3) and Remote a 32K CPU can support only two terminals and two satellites concurrently. Only one user at a time can operate in background (compilation/assembly). By comparison the Forth implementation supports three terminals, four satellites and host-host interchange. No restriction on compilation/assembly exists. The complex, tedious, and time consuming linkage steps required under a standard operating system disappear in Forth. Installation of real-time tasks involves a few milliseconds of time and reconfiguring the network is a matter of 2-3 minutes - not hours.

The system is ideal for a network where satellite processors perform the function of data collection for 1-3 instruments, and where it is essential that maximum throughput exist in the satellite with minimum memory requirements. The network permits software development at remote terminals (virtual terminal operation) on satellites while real-time data collection continues without interruption. The software is procedure oriented. It involves use of an agglutinative back-linked dictionary, which permits the development of code that is far more human oriented than Basic or Fortran. In the satellite neither the speed limitation of Basic nor the memory requirements of Fortran are encountered.

The satellites process the acquired data, then send concentrated information to the selected host for storage and correlation. All of the CPUs, host or satellite, support ordered index file management, a report generator, and other facilities necessary to the generation of good laboratory reports.

FORTH - A LANGUAGE, A SYSTEM, A NETWORK

Forth can be called a computer language, an operating system, an interactive compiler, a data structure, or a virtual computer emulator depending upon the viewpoint from which it is seen. Charles Moore developed the approach while at the National Radio Observatory nearly a decade ago[1]. It has become the most omnipresent software force in astronomy. Forth is in use at the Hale, Steward, Kitt Peak, Owens Valley, Jodrill Bank, Montana, Canberra, and Harvard-Smithsonian observatories. Users outside of astronomy include NASA, Jet Propulsion Laboratories, Princeton Applied Research, RCA Sarnoff Laboratories, the Harvard/MIT Biomedical Engineering Center for Clinical Instrumentation, Chrysler, Mount Sinai Hospital, Western Union, Hewlett Packard, and the Department of the Navy.

The following is presented so that a relatively experienced programmer, unfamiliar with Forth, may gather enough of the flavor of language to enable him to begin reading program listings. It also provides a philosophical background about Forth, and its use in implementing a network.

COMPUTER LANGUAGES - FOUR GENERATIONS

There seems little agreement as to what Forth is. In the days when it was common to discuss computer hardware generations, its author described it as a fourth generation computer language and named it accordingly[2].

Computer software development was closely related to hardware capabilities prior to the recent genera of computers. First generation languages are those spoken only by machines. That is, they comprise the bit pattern sequences which are decoded as instructions by computer hardware. These are also called machine languages. Early machines were programmed only via machine language.

The coming of the second hardware generation brought with it assembler programs and the first compiler languages. An assembler converts mnemonics for computer instructions and user defined symbols into machine instructions. A line of assembly language source is converted into from one to three machine instructions depending on the addressing structure of the computer for which the machine code is generated.

A compiler is a program that converts source for a high level language into either assembly language or a special intermediate code. This allows programmer use of powerful high level instructions which require several machine language execution steps to accomplish. Early versions of Algol, Fortran, and Cobol fall into this category.

Third generation machines were fitted with large mass storage capabilities running software designed to execute in batch environments. Capable compilers for Fortran, Algol and Cobol appeared. The first interactive computation through Basic and the first supercompilers, optimized Fortran and PL/1 began appearing. Pascal was conceived during this time.

The drive toward networking, distributed processing, and the interactive environments marks the beginning of the fourth software generation. In order to understand what this transition implies, an explanation of traditional compilers and interpreter languages is required.

The creation of a Fortran program usually involves an editor program to write the source into a file on a mass storage device. When the source is completed by the user, he must submit the file as a whole to a compiler which converts the Fortran source into an assembly language file on the disk. The compiler parses source lines into symbols and actions. The generation of assembly language is then analogous to the "production" step for a Chomsky type 2 model language[3]. The assembly language is then passed to an assembler program which converts the assembly language into relocatable machine code. The relocatable code is then linked to other modules and subroutines with a program called a link loader. If errors are found at any point in this process the user must go back to the edit phase and begin all over again. This is a very time consuming process when large programs are involved. On a minicomputer, compile/assemble/link times in excess of an hour are not unrealistic. Where programming is done by iteration involving an asymptotic approach to perfect program function, any system which can reduce program correction iteration time will

speed software development.
   This philosophy motivated the development of the Basic language, the traditional interactive interpreter. It is interactive in that source is entered through a keyboard and scanned for syntax errors as each line is completed. This helps remove one of the most common error sources. Debugging facilities are made available which allow program execution control and the examination of variables in an interactive fashion. Source text is converted to execution code on a line by line basis. Every time a high level source line is executed it must be interpreted. In a loop structure, for example, each line within the bounds of the loop is interpreted for every iteration of the loop. This is a very slow process.
   The compiler approach is time intensive during software generation whereas the interpreter allows an interactive approach which reduces development time. At execution however, the speed advantage is strongly toward the compiler languages. Some Basic language processors now contain compile functions in order to ease this liability.

FORTH AS AN INTERACTIVE COMPILER

   Given the strengths and weaknesses of compilers and interpreters, a new approach to high level language processing would combine the strengths of the two traditional approaches. The new approach should be interactive and generate compiled code.
   Forth runs in two modes. These are the compile or defining mode and the execution mode. All source, be it input from the console or from a mass storage device, is handled exactly the same way. Source for definitions is converted to a data structure which allows identical processing of all high level code. This data structure no longer contains the source but is rather a list of execution addresses of the routines that must be executed. Program execution control is available from the console, as is the ability to examine user defined variables.

FORTH AS A DATA STRUCTURE

   The data structure that comprises a Forth operating system can be divided into three functional parts. These are the dictionary, the outer interpreter, and the inner interpreter.
   The dictionary is a series of program entries, built in a special structure, which contains all of the routines or

programs available to a user. This list is linked together so
that it may be searched for a program name beginning at the most
recently defined entry. Figure 3.1 shows the dictionary location in a Forth memory map. Figure 3.2 shows how the dictionaries are linked to the basic Forth dictionary, and Figure 3.3
shows the structure of a dictionary entry. This structure
varies with the actual implementation of Forth and may be somewhat more complex in systems designed to speed the loading of
new programs. The actual contents of the dictionary is dependent upon what source a user has loaded from either the console
or disk.

The outer interpreter has one of two functions depending
on the mode that is active. In the execute mode, a single word
is parsed from the input buffer and a copy of that character
string is compressed or truncated (depending on the implementation) into the form that would appear in the first words of a
definition if it were a definition. The dictionary string is
then searched for an identical header entry. Should it be
found, the routine is executed. If it is not found in the
dictionary, an attempt is made to convert the character string
parsed from the input buffer into a number. This number is
then placed on a user parameter stack where all arithmetic and
logic operations occur (Figure 3.1). Should the string not
represent a number, an error routine is invoked to inform the
user.

The outer interpreter has a different function if it is
operating in the compile mode. This mode is entered by the
execution of a definition whose purpose it is to create new
definitions. The two most common of these are ":" and "CODE".
The colon character begins all high level definitions and the
word "CODE" begins all assembly language definitions. The
execution of these two markers causes the construction of a
definition header in the dictionary for the new definitions.
The code pointer in the header is filled differently depending
on the definition type. For all definitions which begin with a
colon character, called "COLON definitions", the code pointer
points at the inner interpreter. The code pointer of a "CODE
definition" points at the first word of the parameter field
which will contain executable assembly language when that type
of definition is completed (Figure 3.3).

The construction of a colon definition continues with the
parsing of character strings from the input buffer. If a string
represents a previously created definition name, the address of
the code pointer of that definition will be placed in the definition being created and then the next new string is parsed from
the input buffer. If the string represents a number, the number

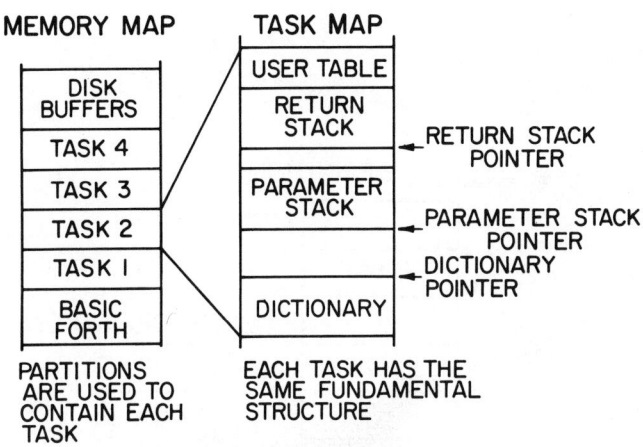

FIGURE 3.1. **MEMORY UTILIZATION**

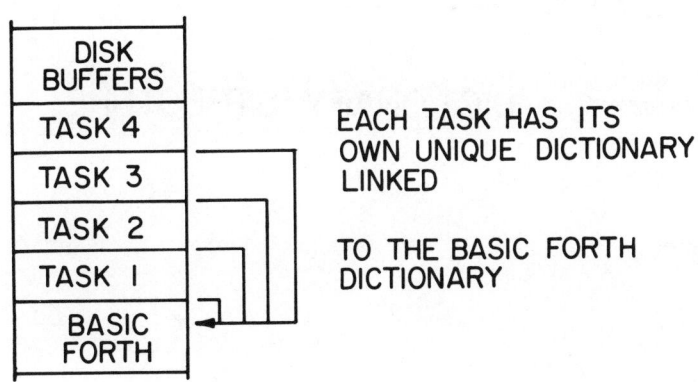

FIGURE 3.2. **DICTIONARY LINKING**

47

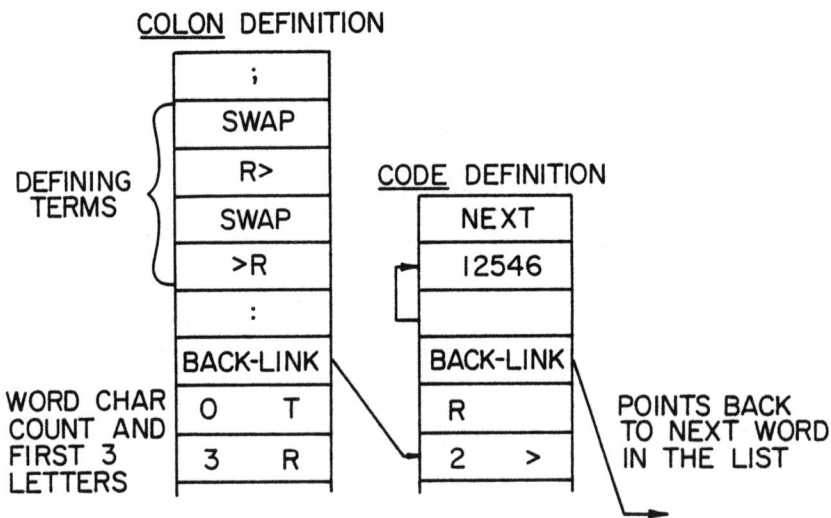

FIGURE 3.3. **DICTIONARY STRUCTURE**

intepreter will place a call (an address) to a routine called
LIT followed by the value for the number. At execution time
the purpose of *LIT* is to place the number following the call on
the user parameter stack. This parameter stack is used to pass
values between routines.

There must be a mechanism to allow exit from the compile
mode. There is a precedence bit in the header for a definition
that allows this. If the precedence bit of a definition is set
the outer interpreter is forced to the executive state and will
cause execution of the routine regardless of the state that
Forth is running in. The word ":" clears the precedence bit
and forces compilation. The routines which terminate compile
mode activity have their precedence bits set. The definition
which terminates the construction of a colon definition is the
";" character. When it is found in the input stream the definition being built is terminated and the state of the outer interpreter reset to the execution mode.

Thus Forth has the ability to locate and execute programs
as well as extend its dictionary. Note that this obviates the
need of the linking loader step that is found in the Fortran
compile sequence.

FORTH AS A VIRTUAL COMPUTER EMULATOR

The inner interpreter of Forth is also called the colon
processor. It is the software that actually executes the string
of addresses that is presented to it in the form of a colon
definition. Colon definitions may contain the names of other
colon definitions; thus the colon processor itself must obvious-
ly be reentrant. That is, it can be called again prior to completion of execution. This means it must include a stacking
function to save the machine state before it starts processing a
definition. This information is stored on the return stack.
The return from a stacked state is also provided by the execu-
tion-time function of the ";".

It is the operation of the colon processor that can be
compared to the hardware with which a computer executes machine
language programs. The inner interpreter must keep track of
which address is being processed in the string of addresses
which constitutes a colon definition. This is done by a
register called the "Instruction counter" (IC or I). The IC
points at the address of the place where the address of the code
pointer for the definition to be executed is stored. As a call
to a definition within a definition is finished, the instruction
counter is bumped to point at the next address. This is remark-

ably similar to the function of the program counter (PC) in most computers. The PC contains the address of the next memory word to be executed. Hardware uses this register to fetch the contents of that address for decoding into a series of actions sometimes called *microcode* steps. The function of the IC is to supply the address of the next *macrocode* step to software which in turn decodes the contents of the place the IC points into a series of actions. Thus Forth can be viewed as a virtual computer emulator capable of having very powerful compound instructions added to its instruction set. A typical register utilization (for PDP-11s) is shown in Figure 3.4.

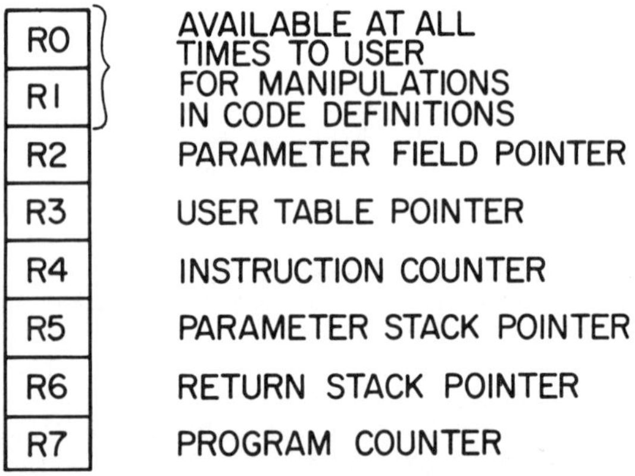

R0	AVAILABLE AT ALL TIMES TO USER FOR MANIPULATIONS IN CODE DEFINITIONS
R1	
R2	PARAMETER FIELD POINTER
R3	USER TABLE POINTER
R4	INSTRUCTION COUNTER
R5	PARAMETER STACK POINTER
R6	RETURN STACK POINTER
R7	PROGRAM COUNTER

FIGURE 3.4. **REGISTER UTILIZATION**

A USER'S INTRODUCTION

The most obvious feature of Forth is its great dependency on a parameter, or user, stack. A stack is a contiguous area of memory which is used to temporarily store data. This is located in high memory and builds downward toward the dictionary which builds upward (Figure 3.1). When the dictionary and the stack meet the computer's memory is full. As shown in Figure 3.4, a stack pointer is kept which points to the next memory word to which data can be written. On PDP-11 systems, the *pushing* of data on the stack involves decrementing the pointer and placing the data where the stack pointer now *points*. *Popping* a value from the stack requires retrieving the data from the location currently specified by the stack pointer and then incrementing the pointer.

STACK OPERATORS

All of the normally required stack operations are implemented in the fundamental part of Forth called the kernel. All numbers which are typed on the console are automatically converted from character strings and placed on the stack. Manipulation of numbers already on the stack may be done using any of the following operators:

SWAP Switches the order of the last two numbers placed on the stack.

DROP Destructively drops the last element of the stack.

OVER Places a copy of the value one inside the stack on to the top of the stack.

DUP Copies the last element of the stack on to the top of the stack.

ROT Places the element that was third on stack on the top while pushing the first and second elements downward. (Figure 3.5)

-ROT Places the top element of the stack where the third element was and floats the old second and third upwards.

PICK   Copies the n th element of the stack
            to the top.

ROLL   Places the n th element of the stack
            on the top while pushing overlaying
            elements downward.

PARAMETER STACK

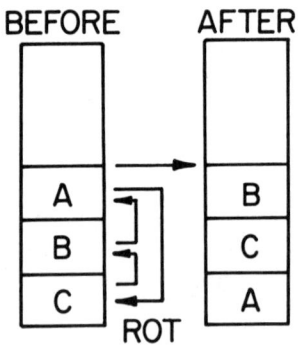

FIGURE 3.5.   **EXECUTION OF ROT**

The following line of executed Forth code places numbers on the stack and manipulates them:

    1 2 3 4   ROT   SWAP   DROP

It is often necessary for the new Forth programmer to write "stack maps" of his code so that he can "see" what is really happening. A stack map of the code line would be as follows:

code	1	2	3	4	ROT	SWAP	DROP
stack →	1	1	1	1	1	1	1
	→ 2	2	2	2	3	3	3
		→ 3	3	4	2	→ 2
			→ 4	→ 2	→ 4		

Stack pointer (→) always points to the top of the stack. It moves down as items are *pushed* onto stack and up as items are *popped* off the stack. This stack orientation, together with the interpretive nature of the approach, led to Sach's name for his implementation of Forth[4]. He called it STOIC for "Stack Oriented Interpretive Compiler."

MATHEMATICS

A ramification of this stack orientation, and indeed, an example of the Forth philosophy, is that mathematics is required to be expressed in Lobachevski (Polish or postfix) notation. That is, the operation of multiplying four times five is expressed as 4 5 * rather than the usual 4 * 5. This is the same *operand operand operator* format used by Hewlett-Packard calculators. The reason for the retention of this structural feature is that it simplifies the construction of the compiler and speeds its execution. A Fortran compiler must convert mathematical expressions from the usual infix notation to postfix prior to the generation of execution code in order to assure the most efficient utilization of the computer's registers. The usual algorithm for doing this involves two stacks, one for operators and one for operand symbols, and the formation of a tree of symbols which is used to guide the code generation step. Operator precedence must be checked with the finding of each new operator as the input line is parsed, and the explicit precedence denoted by parenthesis must be resolved. This can be quite a time consuming task. The driving force in the development of Forth has been simplicity. The programmer is allowed to do what reduction of mathematical complexity he wants to do but he is then required to present the mathematics to be executed in postfix. A little programmer work is traded for a significant simplification of the compiler. This also allows math to be executed directly from the console in the interactive mode with no changes in syntax from that which would be

54   FORTH, A Fourth Generation Language System

used in a program or definition.
   The usual mathematics operations are available. These include:

+   Add the top two elements of the stack leaving the answer on the stack. (Figure 3.6)

-     Subtract the value on top of the stack from the value under the stack.

*   Multiply the top two elements of the stack.

/   Divide the value on top of the stack into the value under the stack.

/MOD   Divide the value on the stack into the value under the stack and return both the result and the remainder.

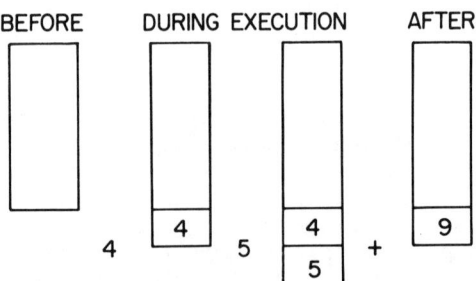

FIGURE 3.6.   **EXECUTION OF 4 5 +**

Similar operators for utilizing 2 word integers (32 bits) are available. Depending on the Forth system their names begin with either a *D* or a *2* .

Many other operators have been added to various versions of Forth and these tend to be implementation dependent. These range from floating point, array, matrix, and transcendental math to Fourier transform packages.

VARIABLES AND STORAGE OPERATORS

User variables can be of three basic types. The first is the single valued constant. The Forth word *CONSTANT* is a defining word which is used to generate a named number. When a constant name is invoked one wants a value and not an address of a place where data is stored. The following line of code will create a definition for *PI:*

31416 CONSTANT PI

The definition *PI*, when executed, will put the value 31416 on the stack.

There is also need for single valued variables. That is, operations which will place the address of a place where data is stored on the parameter stack. The use of the address rather than the data itself is desirable because one may want to either fetch or store data from that place or use the data held in that place as a pointer to a new storage area. The creation of a named variable is straightforward. The following line creates a variable called *AREA* whose initial value is zero:

0 VARIABLE AREA

invoking *AREA* returns the storage address of that variable to the stack.

It is usual that variables so defined will be allocated 16-bits, or one word, on a 16-bit computer. Greater precision may be retained in calculations through the use of double precision integers, those that will use two words of memory or 32-bits. The words *DINTEGER* or *2VARIABLE* provide the defining function for such variables. Floating point variables, in those systems which have been extended to include float math operators, are usually defined through the same word as 32-bit integers because they also require 32-bits per stored value. This should indicate to the observant that Forth does not check data type during math operations. This is indeed true. Such error checking is left to the programmer to include in those places where he will rationally need it.

The third type of variables that one might want to utilize are those that are multiply valued, i.e., data arrays. The word *ALLOT* is used in conjunction with *VARIABLE* to create data arrays. The following creates an array called *DATA* which has 1000 elements available to it:

0 VARIABLE DATA 999 ALLOT

The execution of *DATA* will place the address of the first element of that array on the parameter stack.

One now needs the ability to use the addresses provided by the execution of words defined by *VARIABLE*. The @ operator, pronounced *fetch*, is used to replace the address on the stack with the value or data held at that place in the computer's memory. The "!" operator, pronounced *store*, is used to place the data one under the top of the stack at the address held in the top element on the stack. Thus, the following would place the value held in the first element of the array *DATA* on the stack:

DATA @

In order to sum the first two elements of *DATA* and save it in another variable called *SUM* the following could be executed:

DATA @   DATA 1 + @   +   SUM   !

The use of addresses rather than values in this case appears to complicate the matter but is the most general possible solution to the addressing problem. In more complex cases other languages require the use of special data types in order to meet the need for pointer variables. Any variable can be used as pointer variable in Forth if it is needed by the programmer. Should the data in a variable be a pointer to data, a second fetch will rationalize the address. If, for instance, the array *INDEX* contains a list of entry pointers to a very large array *DATA* comprised of individual records, then the following would yield the first data point for the third record:

INDEX   3 +   @   DATA   +   @

This structure is called an indexed excess method (IAM) and is trivial to implement because Forth is inherently pointer oriented. The use of such pointer structures is extremely important in the generation of data base management systems and as a result such systems are easily implemented with this language.

DEFINITION TYPES

A program entry which has a name attached to it is called a definition. The search list of all definitions known to a Forth system at the time of the search is called the dictionary. A group of definitions which have generally related functions is called a vocabulary.

A definition can be of two basic types. The first is called a *COLON* or high level definition from the colon used to begin all such definitions. All definitions known to a system may be used in a colon definition except those used for the generation of assembly language code. The ":" corresponds to the *BEGIN PROCEDURE* of Algol or the *PROCEDURE* of PL/1. Colon definitions are terminated with a ";". This denotes the *END* used in both Algol and PL/1.

The second type of definition is a *CODE* definition. It begins with the word *CODE* followed by a name and contains assembly language for the computer upon which it is to be executed. Movement of Forth programs from one machine to another involves translation of all code definitions if the machines have different Assembly languages.

The ability to intertwine Assembly language with high level code is an essential capability in a language which is to be used in a real-time environment. In the data acquisition world, it is commonly necessary to do extensive I/O programming for purposes for which vendors make no provision in their software. Unusual experimental constraints such as high speed data acquisition or the overlapping of data acquisition with refresh of displays or disk transfers may require that certain portions of the software execute as fast as possible. Compute bound tasks may be optimized for execution speed through the selective use of assembly language. A common approach is to search out the 20% of the code that requires 80% of the execution time and rewrite that portion in assembly language. Such a step may be employed with Fortran oriented systems, but requires linking the Fortran and the Assembler object modules. Forth systems handle ":" and *CODE* definitions concurrently.

FORTH AS A STRUCTURED LANGUAGE

Forth is a structured language in that *functional blocks* or procedures are invoked by name in other routines. The traditional *block structured* compilers are PL/1 and Algol, with Pascal as a relative newcomer. These languages are *top down* structured. That is, driver programs can be written into the source before the functional units of the program are completed. A *bottom up* structured approach is one in which the subunits are written first and then the driver is written.

When a colon definition is compiled by a Forth system, each word it uses is sought in the dictionary. Should it not be present, an error message is generated. This is characteristic of single pass compilers. In a two pass compiler, the first pass builds a table of all used subunit, subprogram, or variable names. The second pass then uses the table to check the *legality* of such names as source is converted to the assembly language or interim code required to implement the high level source.

It would be natural to assume that Forth's single pass structure strictly restricts it to a *bottom up* approach as no forward references are allowed. In practice, however, code is (or should be) written in a *top down* manner and then entered *bottom up*. This is not as complex as it sounds because of the great modularity of the code.

CONTROL STRUCTURES

All computer languages must provide control structures to allow decision making which can alter program flow or allow iteration on certain sections of code.

The most common control structure in Fortran and Basic is the *GOTO* statement. This directs program flow to a line of source code denoted by a label number. *GOTO* statements are combined with decision statements such as *IFs* to direct program flow on a contingency basis. An example of a condition flag decoder written in Fortran can be seen in Figure 3.7.

As is the case with most structured languages, Forth lacks all line numbers and as a result the *GOTO* is not used. In Figure 3.8 the Forth required to implement the same command decoder is seen. It is shorter in terms of source length and requires very few words of memory (18) as a nested series of *IF ... ELSE ... THEN* structures. In such a structure the clause between the *IF* and the *ELSE* is executed if the test is true and the clause between the *ELSE* and the *THEN* is executed if the test

```
C       FORTRAN CONDITION FLAG DECODER
        IF (IFLAG. EQ. 1) GOTO 1000
        IF (IFLAG. EQ. 200) GOTO 2000
        IF (IFLAG. EQ. 400) GOTO 3000
        GOTO 4000
100     CONTINUE
          .
          .
          .
C       SERVICE ROUTINE FOR IFLAG EQUAL 1
1000    CONTINUE
          .
          .
          .
        GOTO 100
C       SERVICE ROUTINE FOR IFLAG EQUAL 200
2000    CONTINUE
          .
          .
          .
        GOTO 100
C       SERVICE ROUTINE FOR IFLAG EQUAL 400
3000    CONTINUE
          .
          .
          .
        GOTO 100
```

FIGURE 3.7. FORTRAN *Condition Flag Decoder*

```
        : 1-SERVICE ... ;
        : 200-SERVICE ... ;
        : 400-SERVICE ... ;
  : SERVICE IFLAG @ DUP 1 = IF 1-SERVICE ELSE
        DUP 200 = IF 200-SERVICE ELSE
        400 = IF 400-SERVICE
        THEN THEN THEN;
```

FIGURE 3.8. FORTH *Condition Flag Decoder*

condition was false. The flow of execution then continues for both cases at *THEN* in the line of code.

Similar structures are found in Algol, PL/1, and Pascal. In the 1960 standard Algol the usual structure is *IF THEN* ... *ELSE* ... . The problem with this form is that the *THEN* refers to the true clause. The reuniting of program flow occurs at the end of the *ELSE* clause but there was no marker indicating where that really occurred. The 1968 Algol standard added the keyword *ENDIF* to provide that marker. The Forth form eliminates the need for the *ENDIF* while retaining the basic Backus Naur Form (BNF) which has been found so useful.

Many loop control structures are available in Forth. Among these are:

>   BEGIN ... <test> END
>   BEGIN ... <test> WHILE ... AGAIN
>   DO ... <test> WHILE
>   DO ... <test> UNTIL
>   DO ... LOOP
>   DO ... <increment> +LOOP
>   M N_1 CASE ... ELSE
>   (N_y CASE ... ELSE)$_x$
> N_z CASE ... ELSE THEN (THEN)$_x$ THEN

An additional keyword, *LEAVE* is available within loop structures which forces immediate termination of the iteration. Many of these structures are available in code definitions.

While differing markedly from Fortran, Forth utilizes control structures which are functionally equivalent to many of those available in the most advanced structured languages.

MASS STORAGE AND VIRTUAL MEMORY

A data base management system heavily depends on the use of mass storage devices, such as disks, and large data arrays. Facilities for manipulation of such ancillary storage vary greatly among available minicomputer operating systems. Restrictions on the structure of files are usually present at the operating system level rather than in language compilers. Indeed, there are few standards for this kind of I/O among the compiler languages. Several vendor operating systems contain data base management facilities which are available to their Fortran compilers. Examples of these are Digital Equipment Corporation's DMS, Data General's INFOS, and Hewlett Packard's IMAGE. Unfortunately, these facilities are usually available only for the vendor's larger or more advanced computers (e.g., the PDP 11/34 through the VAX 11/780, and DG's Eclipse computer line). An additional problem is that many of the operating systems under which these facilities exist are slow as a result of providing comprehensive inter-user protections.

Pascal is unusual in that data structure definition is encompassed within the variable defining capability of the language. Forth has similar capabilities. It is ideal for the typical record management systems needed at the instrument application level.

No file structure is imposed on the systems architect by Forth. This can be viewed as a disadvantage if one is not willing to prepare his own approach to data structures and mass storage allocation. Its great strength is that a programmer is in no way inhibited in designing fast, efficient algorithms which suit his needs rather than attempt to meet the hypothetical general need. It also simplifies network architecture as will be shown.

The basic unit of data on a Forth mass storage device is called a *block* and consists of 1024 bytes. They are numbered sequentially from zero. Certain areas of the primary disk on a Forth system are usually allocated to the common requirements of all Forth systems. It is usual to locate the bootstrap loader for the disk subsystem in block 0 with a pre-compiled image of the Forth kernel in the next few blocks. The next large area of disk is reserved for the source for systems programs as the rest of the Forth system is literally compiled from source during system startup. The next area is that reserved for applications programs and the storage of data for users applications.

The primary vocabulary word provided for the manipulation of disk data is *BLOCK*. One places a number on the stack, executes *BLOCK*, and it returns the address of the first word of memory now containing the requested data. *BLOCK* scans the allotted disk-buffer areas to ascertain if the requested block is memory resident. If so, the address of that disk-buffer is returned. If not, a disk-read is implemented. Virtual memory is an inherent aspect of Forth. Should one wish to return data which has been changed or up-dated to the disk, the word *UPDATE* marks the buffer so that *BLOCK* will cause its write-back if it needs the buffer. One can force the immediate write-back of data through the use of *FLUSH* after an *UPDATE*. The number of system disk buffers available to Forth depends upon implementation with two to seven usually being provided. This approach makes implementation of virtual array handlers quite simple.

Very large data arrays represent a potential difficulty for the user of most compiler languages, with the notable exception of Pascal. Forth can provide for such large arrays. Several implementations of virtual arrays (those actually residing on mass storage rather than in memory) exist. One uses a 32-bit address pointing to a word on the disk. A variable which has been defined as a virtual array places the 32-bit address on the stack. The virtual operators (*V@* and *V!* which are analogous in function to "@" and "!") then use this base address and ancillary offset arguments in conjunction with *BLOCK* and *UPDATE* to provide for fetching the block of disk data containing the desired value, the calculation of the address in real memory, and the capability to write-back the data to disk. Perhaps 30 lines of code, and a few words of programming implement virtual arrays.

RELOCATION, OVERLAYS, JUMP TABLES, ALIASES, AND STRINGS

With this background it is possible to focus on several other features of Forth that are usually available in classical operating systems only at a very high price in time, space or money.
Entries in the Forth dictionary can be laid down directly in any available space. This avoids the relocation problems inherent in relocatable code. The structure of the language, both at *COLON* and at *CODE* level, makes preparation of such position independent code simple.
Since no linker is involved overlay execution is simplified. With a back-linked dictionary it is possible to define words such as *XXXX FORGET*, which will scan the dictionary until the indicated word is found, and then reset the dictionary pointer to the end of the previously defined word. Overlays may begin at this point. Multiple fence boundries may be created in this way. Memory can be used very efficiently as successive multiple overlays are invoked.
Although Forth is best employed in a strictly structured programming mode, it is possible to develop jump-tables in which a list of previously defined Forth words are arranged in a table. The word invoking the jump operation calculates an offset into the table, picking up the area to which control should be passed. When the required function is completed, control returns to the instructions following the word which invoked the jump. This is really a jump-to-subroutine-table operation.
The alias function may be implemented by defining a dummy word. During execution of the dummy word the execution of another Forth word will occur by aliasing the dummy operation to the previously defined Forth word.

Multitasking

String operations perform functions identical to *DUP*, *SWAP*, *ROT*, *OVER*, and *DROP*. They act on alphanumeric strings stored on a separate string stack and are easily implemented in Forth. Translational words move ASCII strings representing numbers from the string stack to the parameter stack, or vice versa.

MULTIUSERS

In a typical DEC PDP-11 installation, register use is optimized, as shown in Figure 3.4. Registers 0 and 1 are always available for user use. Register 2 always points to the parameter field of the currently executing instruction, and is useful to carry arguments into *CODE* instructions. Register 4 is the instruction counter, Register 5 the parameter stack pointer, Register 6 the return stack pointer, and Register 7 the program counter.

Register 3, called *U* or *USER*, is the key to much of Forth's power. It always points to the area in memory specific to the task currently being executed. *USER* is actually a large table (20 to 40 numbers) of information vital to that task. Task-specific variables and constants are stored in *USER*. As indicated in Figure 3.2, each task has its own dictionary area where words specific to its function are laid down. Each task also has its own *PARAMETER STACK* and *RETURN STACK*. Bottom and top positions of these stacks can be recorded in the *USER* table. To switch context of operation fully from one task to another, all one needs to do is store the values of the *INSTRUCTION COUNTER*, the *RETURN STACK POINTER*, and the *PARAMETER STACK POINTER* via the current *USER* table, and reload these values via the next task's *USER* table.

MULTITASKING

Multitasking is essential in most laboratory systems. The implementation discussed here is based on the FORTH, Inc. version. How is control passed from one task to another in an orderly manner? Again, the key is the *USER* table. It contains two variables called *STATUS* and *ADDRESS*. They help Forth implement tasks much like hardware implements direct-memory-access (DMA) in a computer (Figure 3.9).

The DMA concept recognizes that most tasks involve the transfer of a predetermined number of data points to and from memory. A *WORD-COUNT* register is set up, indicating the

64   FORTH, A Fourth Generation Language System

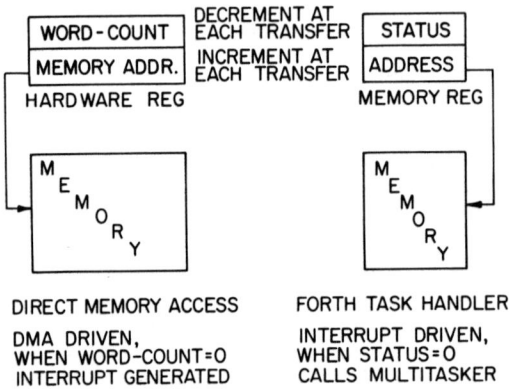

FIGURE 3.9. DIRECT MEMORY ACCESS VS FORTH HANDLER

number of words. The contents of the *MEMORY-ADDRESS* register knows where the initial location data are to be transferred to and from. Each time a word is transferred, the DMA hardware steals time (cycles) from the computer. The *MEMORY-ADDRESS* is incremented, the *WORD-COUNT* register decremented, and a data word is transferred. When the *WORD-COUNT* register goes to zero, an interrupt to the computer signifies the transfer is complete, and appropriate action is taken by the interrupt-handling software.

When a Forth task wishes to send and receive data, the Forth multitasker uses the *STATUS* word as a *WORD-COUNT*. Positive values indicate it is going to transmit, negative values that it is going to receive. *ADDRESS* contains the initial memory location involved. Each time a word is transferred, an

interrupt to the computer is involved. *STATUS* is incremented
or decremented appropriately, *ADDRESS* is incremented, and a
data word is transferred. When *STATUS* goes to zero this signifies that the transfer is complete.

Forth has a round-robin scanner, shown in Figure 3.10
that continuously monitors the value of *STATUS* for all the
tasks installed on the system. If the value of *STATUS* is nonzero, the Forth multitasker ignores it, proceeding around the
circular linked list comprising its known tasks. If the
value of *STATUS* is zero, the multitasker, using the values for
pointers stored in the *USER* table, begins execution of the
next Forth word associated with the task. Installation of a
new task conceptually involves momentarily breaking the circular-linked list and placing new forward pointers in two
locations. The time needed to install and remove tasks is

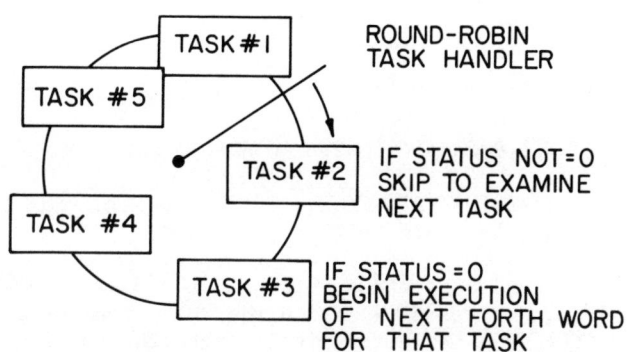

FIGURE 3.10. **MULTI TASKER OPERATION**

minimized. In addition, the round-robin task handler is
exceedingly fast. A 16-channel multiplexed A/D converter
handled by an A/D task handler in Forth, which permits each
channel to have independently controlled time, rate of data
acquisition, data buffer area, and number of points to be acquired, has been benchmarked at 120 points/sec/channel (2 KHz
total throughput) while supporting a 30 KHz scope display.

DUMMY TASKS

   The VPI&SU network demonstrates the ease with which
tasks can be created and installed on the system. As each host
or satellite in the net is brought into service six dummy tasks
are installed on each node. These are given names such as
OTASK, 1TASK...5TASK. A word called *TERMINAL* performs this
installation, which allocates space for six task stacks in
memory (Figure 3.1). During system-build time these tasks,
OTASK...5TASK, are not installed in the round-robin. They are
called dummy tasks since they exist only to allot organization-
al space for future task consignment.
   A task may be consigned to a dummy task space by the code:

   : *ADC-TASK TASK* ;

   CONSIGN ADC-TASK

   ADC-TASK ADC-VECTOR CONFIGURE-IN

   : *AD-TASK ADC-TASK ACTIVITY AD-DATA #POINTS DMA =
      WAIT FINISH* ;

TASK aliases a dummy task to the user name *ADC-TASK*.
CONSIGN places *ADC-TASK* in the round-robin. *CONFIGURE-IN*
and *CONFIGURE-OUT* automatically make the interrupt vector
assignments for interrupt driven data transfers. *ACTIVATE*
starts execution of the task at the words following activate.
Task-builders on many systems measure installation times in
minutes. The method described above does it in milliseconds.
   These tasks can run in real-time collecting data or
monitoring I/O control/sense registers. They may call other
tasks defined in a similar way. They may share the screen of
a *dumb* CRT employing split-screen mode, or take over the terminal task when full reports are available or requested.

Since each task, so defined, takes only the buffer space allocated to data collection, and a minimal return stack, installation of many tasks on the system poses no problem.

NETWORKING

All Forth disk transfers involve a call to a routine called *BLOCK* which supervises transfers and insulates the user from the idiosyncrasies of various disk subsystems. It is a relatively simple matter to write *BLOCK* to transfer requests and data to and from a special satellite task in a host processor via a serial line between the host and satellite. The network protocol is considerably simplified in this situation. The host is always waiting for requests from the satellite which consist of a two byte block number incorporating a read/write selection bit, followed by a checksum byte. On a read request, the host obtains the block requested and transmits it to the satellite. On a write request, the host waits for reception of the 1024 bytes making up a block and then writes these bytes to disk, using its own copy of *BLOCK* A checksum byte is appended to every transfer to assure data integrity.

The block buffers that exist at the host and satellite are identical in form and use. The simple transfer protocol involved means that it is possible for satellites to have satellites. Since the multiuser terminal handler is the same in host and satellite locations it is possible to have multiple terminals on satellites.

In summary, the block transfer nature of Forth leads to a very simple network protocol and to a very flexible network structure.

In operation most of the Forth compiled on a host is compiled into each satellite as it is brought on line. The satellite does not run a simulated OS. Other networks which require such simulation often also require subtle and enigmatic program changes to run copies of programs in the two environments - host or satellite. With Forth it is also possible to have personalized load modules which create individual Forth kernels at each satellite if that is desirable.

The virtual terminal capabilities of the satellites allow even more application-specific programming to take place. Since the satellites are responsible for only a small number of tasks and have minimum software overhead they can collect data very rapidly. Typically, in 16K machines with only 5K taken up by systems programs, the satellites have sufficient space for user programs and data buffering. System intercommunication is via 9600 baud, RS-232 asynchronous lines (3 wires).

The hosts require about 6-8K for systems software, with the remainder allocated to individual terminal dictionaries and satellite buffering. Data and programs can be exchanged between host-and-satellite or host-and-host. *SEND/RECEIVE* and *TRANSMIT/FETCH* equivalences are available for the two routes, allowing either end to initiate a transaction. *GET/RELEASE* functions assure that simultaneous use of a facility by two users is prevented. A *GRAB* command permits usurping control.

Forth is not a competitor for SNA(SDLC/HDLC), X.25, or Decnet (DDCMP). These network protocols are useful in systems involving alternate routing or multidrop transmissions, often via public packet switched facilities. They are extremely flexible, but hardware cost and software overhead times are excessive. Many of the functions they provide are simply not needed at the lower levels of laboratory automation. Most automated instruments need only support point-to-point transmission to a single host used for storage/correlation/retroversion. Parity/check-sum and/or CRC error detecting algorithms are sufficient at this level.

Another implementation for the point-to-point approach is the multidrop IEEE-488 protocol discussed in another chapter. It is more expensive to install since it is a 16-bit parallel bus (8 data lines, 8 control lines).

RECORD MANAGEMENT AND REPORT GENERATION

The presence of an ordered-index file management system and a report program generator recognize that most laboratory applications involve the collection of data in real-time, with subsequent correlation of that information with other data, or abstraction of a sub-set of the data. Ultimately, a neatly organized printed report is needed. Table headings and subheadings, format control, columnar and row control, and pictured Cobol output all make the latter task easy to perform. In ordered-index record management the main records are kept in the order created, containing data such as *SAMPLE ID*, *SUBMITTER*, *SOURCE #*, *RESULT-1....*, *RESULT-2....*, etc. Ordered files of key words such as *SUBMITTER* and *SOURCE* (ordered alphabetically, numerically, or hashed) permit rapid searching for records, and fields within the records, for use by the report generator.

AN APPLICATIONS ORIENTED LANGUAGE

   In the past, the emphasis in scientific articles on computer applications has tended to focus on the hardware architecture of the CPU, interfacing techniques, and, above all, on the low costs involved in this hardware phase. A few articles have stressed that software development costs are of overriding importance, but the interest of the developing computer community in chemistry has been directed toward electronic components. This attitude is hard to justify after a careful cost analysis of operating a micro/minicomputer system is done. In the authors' laboratory, where 12 small computers are used, the cost of the central processing units represents 2% of the total development cost to date. Terminals alone represent 6%, and other peripherals 25%. The single most expensive segment is software development, an estimated 62% and growing.

   Software costs also were disregarded in the development of second generation computers. At that time, this attitude was justified because of the inordinate cost of the hardware and the lack of suitable high-level languages. To a marked extent, the attitudes and practices of that earlier era have been emulated in scientific applications. Two important differences, however, now exist. First, the cost of the personnel to implement software has escalated as hardware costs have plummeted; and second, alternatives in language now exist. The classical languages either sacrifice speed for human cordiality or impose excessive hardware burdens on the systems designer who needs moderate execution speeds *AND* human cordiality. To mix languages by linking object modules is cumbersome. The awkwardness and shortcomings of most languages and their concomitant operating systems stem from the fact that the languages have been created philosophically upon the precepts of large computer systems, extrapolating form and function to the smaller target computers. This results in inefficient use of bulk storage and a voracious waste of time, space, and human energy.

   The need is for a system that permits easy development of a procedure-oriented language. The ultimate goal is to develop, in a task area, those words that will permit a technician to control all of the necessary instrument functions by invoking the proper sequence of names. This is similar to new instruments that permit a series of keypad strokes to generate a program that will operate the instrument in successive steps. The order of operations can be changed by altering the order for the keypad strokes. The Forth system can perform these tasks. At an appropriate point the back-linked dictionary thread can be broken to protect basic names and functions by making them unavailable to the technician. The system will respond only to a limited command set. This makes it difficult to crash. In a few seconds, however, a competent programmer can relink the chain and make corrections, changes, or install new tasks.

Forth systems therefore are characterized by flexibility. Their application is limited only by the imagination of the programmer. This capability makes them truly fourth generation.

Forth stands alone; it is both a language and an operating system. For the first time a software system has grown and developed in the environment where it will be used and for the micro or minimum configuration minicomputers upon which it will operate.

Forth is serving as a greenhouse for the development of new concepts in laboratory and applications software[5].

ACKNOWLEDGMENT

The authors wish to thank Thomas Sargent, Steward Observatory, University of Arizona, H. Wayne Hammond, Jet Propulsion Laboratory, and Dr. Isabel Starling of the author's laboratory for their help in initiating FORTH in this laboratory. Funds from the Gillette Charitable and Education Foundation partially supported the project.

REFERENCES

1. *FORTH Dimensions*, Ed. FORTH Interest Group, *1*, 46 (1980).
2. C. H. Moore, *Astron. Astrophys. Suppl.*, *15*, 497 (1974).
3. N. Chomsky, *Aspects of the Theory of Syntax*, MIT Press, Cambridge, Mass. (1965).
4. J. M. Sachs, *STOIC User's Manual*, Biomedical Engineering Center, MIT, Cambridge, Mass. (1977).
5. A. Taylor, "The Taylor Report: FORTH Becoming a Hothouse for Developing Languages", *Computerworld* (1979).

4

THE GENERAL ELECTRIC
LABORATORY AUTOMATION SYSTEM

W. T. Hatfield, R. P. Goehner, E. Lifshin

INTRODUCTION

The past decade has seen remarkable improvements in the versatility and sophistication of analytical and microstructural characterization instrumentation. Contributing factors have been advances in vacuum technology, increased fundamental understanding of the interaction of various probes with matter, and perhaps most significantly, the electronics revolution. This new generation of instrumentation not only makes it possible to examine more complex materials, like new multicomponent high temperature alloys and ceramics, but also makes it easier to determine the concentrations of trace level components which can significantly affect air and water quality or the behavior of semiconductors and catalysts. Reduced data collection and analysis time as well as the ability to gain information from smaller samples are key factors in improving the productivity of any organization concerned with materials development, quality control or environmental monitoring.

The use of computers of all sizes has been essential to realizing these goals, however in retrospect what now seems to be clear cut justification for the use of integral microprocessors or minicomputers was not always obvious, and, in fact, many of the options available today did not exist just a few years ago. This laboratory, like many others, contributed to the development of concepts which are now standard features in many commercially available instrument systems. The benefits gained from such efforts have been lead time in incorporating computerized capability as well as increased familiarity with this mode of operation. We have also learned, however, that a time comes to abandon one's pet approaches in favor of the more concentrated effort put forth by equipment manufacturers.

The system described in this paper is a constantly changing one which is sufficiently flexible to allow for the addition, deletion and modification of instrument interfaces and the related software. Many issues remain to be fully resolved like unified sample submission procedures and related record keeping, the incorporation of relevant data bases for diffraction and mass spectrometry, and univeral acceptance of the system by all possible users. Nevertheless, we believe that the system described has had the net effect of greatly increasing our overall productivity and the quality of analytical results.

BACKGROUND

The General Electric Laboratory Automation System (LAS) was developed for the Chemical and Structural Analysis Branch (CSAB) at the Corporate Research and Development (CRD) Center. CSAB provides a wide variety of the analytical services for the Center ranging from nuclear magnetic resonance (NMR) to Metallography (Table 1). The Branch is spread throughout the building occupying lab space on different floors and in separate wings, often separated by large distances. The analytical instruments used are of many different vintages ranging from twenty year old diffractometers to the latest type of fourier transform infrared (FT-IR) spectrometer. some of these instruments are in very noisy electrical enviroments which can cause special interfacing problems.

In most cases these instruments can be controlled by using some type of interface which will generate or sense a contact closure, drive a stepper motor, simulate an analog control signal, etc. Newer instrumentation is designed with computers in mind making interfacing more straightforward. Older instruments, on the other hand, are more difficult to interface because of signal incompatability. This problem can often be overcome through the use of devices like shaft encoders to sense position, optically isolated solid state relays and optoisolators. The data produced by these instruments can in most cases be represented by a series of points either in a table or as a spectrum. Once this data is reduced to digital form it can be more accurately analyzed and more easily stored. The challenge then lies in storing data from a wide variety of analytical techniques in a meaningful fashion which will allow them to be most efficiently managed and analyzed.

Table 4.1. Analytical services provided by CSAB.

- ORGANIC SEPARATIONS AND MOLECULAR WEIGHTS
- X-RAY FLUORESCENCE SPECTROSCOPY
- ELEMENTAL ANALYSIS
- ABSORPTION SPECTROSCOPY
- NUCLEAR MAGNETIC RESONANCE
- METALLOGRAPHY AND LIGHT MICROSCOPY
- ELECTRON MICROPROBE
- SCANNING ELECTRON MICROSCOPY
- LOW RESOLUTION MASS SPECTROMETRY
- GAS CHROMATOGRAPHY-MASS SPECTROMETRY
- SURFACE ANALYSIS
- ANALYTICAL ELECTRON MICROSCOPY
- X-RAY DIFFRACTION

In 1972 a system was developed which controlled and acquired data from four instruments in our scanning electron microscopy (SEM), microprobe and X-ray fluorescence laboratory. Part of the rational for this system was that since peripherals such as line printers, mass storage and plotters were not used 100% of the time, they could be shared by the four instruments, thus providing a more cost effective solution than having a separate computer and peripherals for each instrument. A 16-bit minicomputer (an Interdata 70) with 64K Bytes of core memory was used along with a 2.5M Byte disc, cassette tape and line printer. The instrument interfaces were RS-232 compatable, accepting ASCII strings as control commands and returning data as ASCII strings. This allowed control and data acquisition programs to be written in a high level language (FORTRAN) thus speeding up program development and modification (1). This approach also allowed us to take advantage of data reduction programs available in the literature.

The concept was quite successful, however it had a few drawbacks. We found that the memory was occupied most of the time with control and data acquisition programs which ran a long time, therefore data reduction programs which typically ran a very short time had to wait. In addition, the minicomputer operating system was deficient in a number of areas. The file system was rather inefficient and inflexible and it was easy for one user to overwrite other user's files. All programs had to be started from the main system console, since the instruments were all in one lab, this was not too serious a deficiency, but became more important as use of the system expanded beyond one lab. The introduction of the microcomputer offered a solution to one problem if it could provide the control and data collection function thus freeing the central computer for data reduction. It also appeared that many of the deficiencies in the operating system had been overcome in later more sophisticated systems. In 1976 it was decided to expand the scope of the system to one that would be capable of handling a wide variety of instruments. Drawing on our past experience with the SEM - Microprobe system and our experience with microcomputers we were able to define the requirements of the new system.

SYSTEM REQUIREMENTS

1. The system must be easy for the non-computer expert to use.

   Since the system is to be used by a large number of people with varying degrees of experience and expertise with computers, the "human interface" is of utmost importance. Programs must have a question and answer dialog. Diagnostic messages must be clear and concise.

2. The system must be a multi-user system.

   A truly multi-user system is required allowing users to run experiments and analyze data simultaneously from remote terminals rather than the main system console. When using the system, the user must feel that he or she is the only person on the system. The response time between transactions must be kept to a minimum.

3. There must be an adequate file security system.

The file system must prevent one user from accidentally destroying another person's files. There must be a way of easily backing up the disc storage in case of disc failure resulting in loss of data.

4. The system must be able to accommodate both human users and microcomputer interfaces to instrumentation.

   Microcomputers must be able to gain access to the system with a minimum of dialog. The host must be capable of indicating when it is ready to receive data.

5. The system must support on-line program development in high level languages such as FORTRAN and BASIC.

   Program development for a specific application is best handled by the scientist who uses the equipment rather than a computer programmer who is familiar with software, but lacks the detailed understanding of a particular technique. Most of the scientists in our lab are capable of programming in FORTRAN or BASIC.

6. The system must allow communication over large distances and allow remote dial-up capability.

   It must be possible to transfer data to the host from other labs throughout the company. There are many cases in which it is necessary to monitor experiments which run overnight or on weekends. The scientists may wish to do program development from home in the evening.

7. The system must be able to communicate with the CRD Honeywell DPS-2 time sharing computer to transmit data files back and forth.

   There are cases in which it is necessary to transmit data between the two computers to take advantage of particular processing capabilities of one of them.

8. The user must not be dependent on the operation of the host computer for instrument control and data acquisition.

   With the volume of samples and number of instruments and people depending on this system, it is not acceptable to be dependent on the functioning on any one component of the system for its entire operation. Therefore, if the host

76   The GE Laboratory Automation System

computer is down, it must be possible for the individual microcomputers to continue operation.

CONFIGURATION

The LAS consists of a large host minicomputer interfaced to various analytical instruments by means of a series of microcomputer based data acquisition systems (Figure 1). The microcomputers handle control, data acquisition, intermediate storage and any preprocessing of data. The host computer provides archival storage and retrieval of data and processing. In addition, the host provides many of the computational needs of the users.

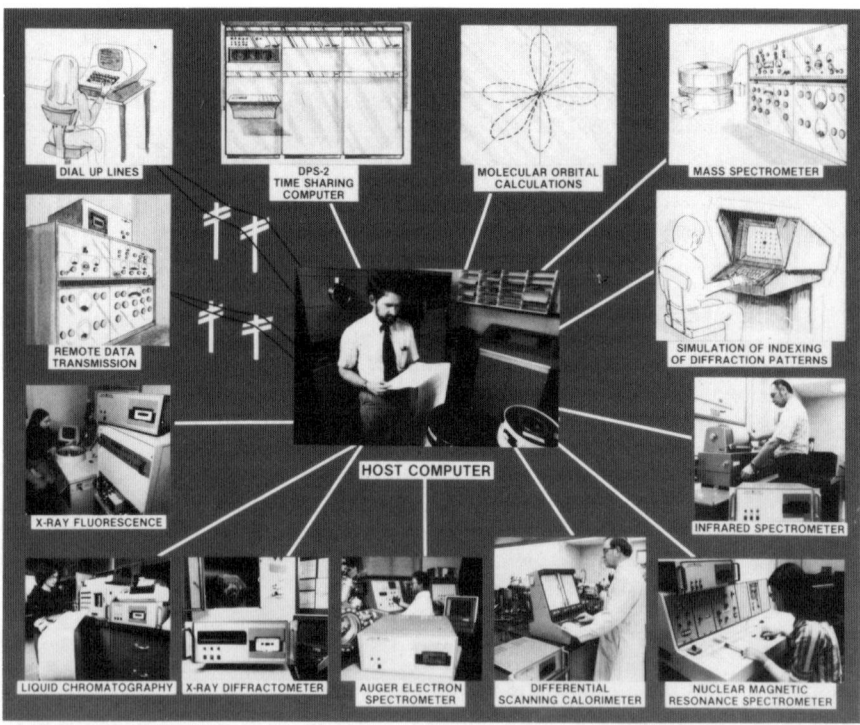

Figure 4.1. General Electric Laboratory Automation System (LAS).

The host computer is a Perkin Elmer 8/32 32-bit minicomputer with one million bytes of 750NS core memory, high performance single and double precision floating point hardware, writable control store (which contains selected FORTRAN run time math routines). Peripherals include a system console CRT, two 67-megabyte discs, a 10-megabyte disc, a 2.5-megabyte disc, a 9-track magnetic tape, a high speed line printer, a Zeta four pen plotter and 3-dial up lines (Figure 2). The microcomputers as well as terminal users communicate with the system over RS-232 compatible Programmable Asynchronous Lines (PAL) at speeds ranging form 110 to 19,200 baud. The system is currently configured for twenty eight lines. The host runs under Perkin Elmer's OS32 Realtime Multitasking operating system with Multi-terminal Monitor (MTM). The operating system supports both a BASIC interpreter and FORTRAN compiler.

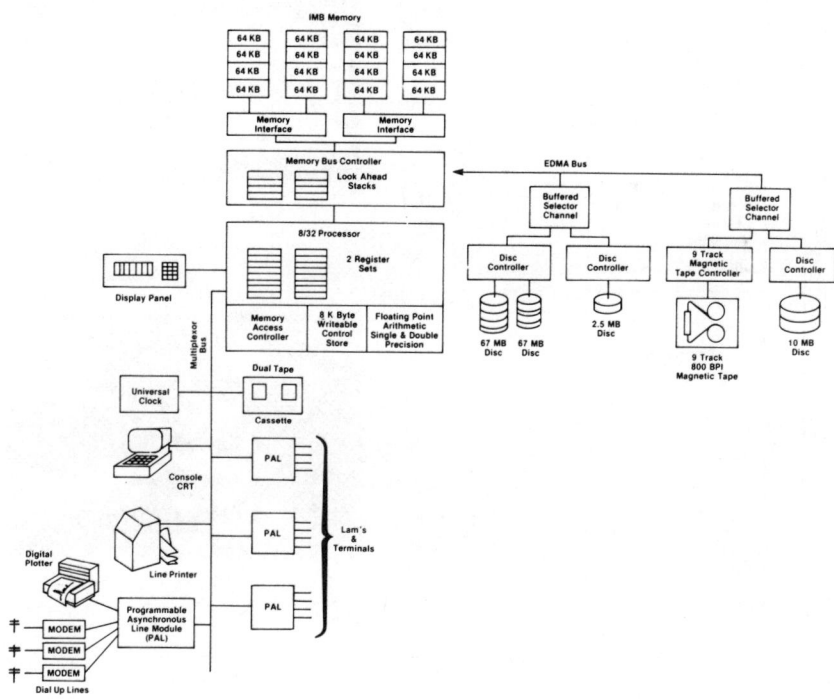

Figure 4.2. Diagram of Laboratory Automation System host computer.

The GE Laboratory Automation System

The microcomputers called Laboratory Automation Modules (LAM) were developed at General Electric. The LAM's use the MULTIBUS backplane, this ensures a second source for many of the board level components. They use 8080's or Z80's as the central processor unit (CPU). The CPU board itself is an Intel SBC-80/20 or equivalent. A BASIC interpreter is provided in read only memory (ROM). This allows easy program development with a minimum of hardware support required. In addition the LAM has up to 44K Bytes of random access memory (RAM), 72 parallel input-output (I/O) lines, 3-serial ports and a cassette tape unit. To this basic module, interfaces are added to tailor the LAM to a particular instrument (Figure 3).

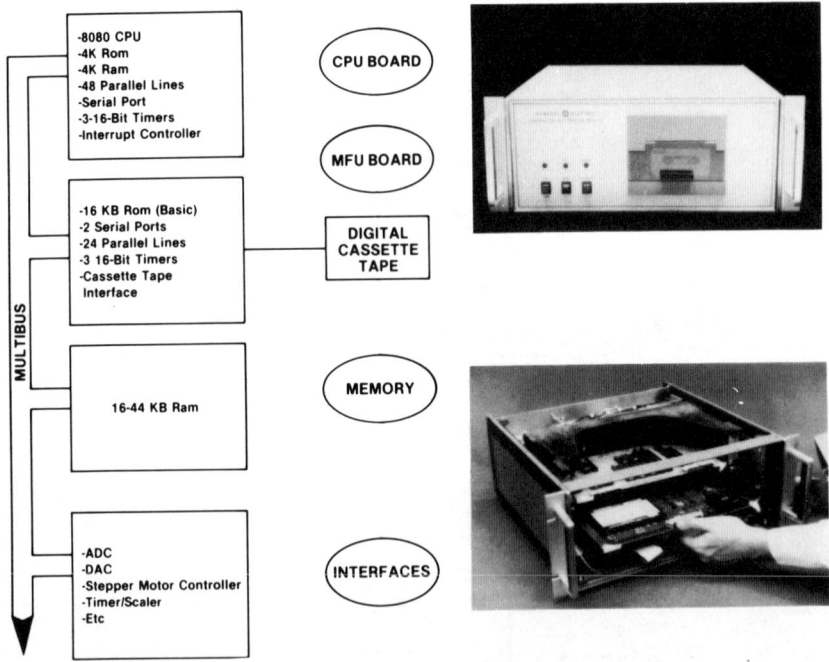

Figure 4.3. *General Electric Laboratory Automation module (LAM).*

INSTRUMENTS INTERFACED

Nuclear Magnetic Resonance (NMR)

The LAM has been interfaced to a Varian T60 NMR using a 12-bit ADC with programmable gain (1, 2, 4, or 8) yielding a 16-bit dynamic range and a 12-bit DAC. Data can be acquired using the internal spectrometer sweep or using the DAC to sweep the spectrometer through the spectrometer's external sweep input. Data can be acquired in the single pass or time averaging mode. The system is used extensively for obtaining accurate peak integrals.

Liquid and Gel Permeation Chromatography (LC GPC)

LAM's are connected to a Waters 150C Gel-permeation chromatograph, a Waters ALC-244 liquid chromatograph equipped with a model 710 WISP auto injector and an Ion chromatograph with conductivity and coulometric detectors. The LAM's sense injection through a bit of a parallel port and in the case of the WISP, can control and interrogate the auto injector through another parallel port. A 12-bit multiplexed ADC is used to collect the data. Each LAM can handle up to eight inputs with variable dead time delays for each input.

X-RAY DIFFRACTION

Two LAM's are used in the x-ray lab. One is interfaced to a Siemens D500 powder diffractometer. It controls the theta and 2-theta motions of the diffractometer and acquires data using a built-in timer/scaler. In addition, it is capable of sensing if the shutter is open and if the goniometer control is turned on. This LAM is equipped with a 4KB ROM operating system and 32-character display and an additional 16kb of ram in addition to the standard features. The display is used to monitor the angular position of the goniometer and the timer/scaler reading. The other LAM is on a Joyce-Loebl microdensitometer. The LAM uses a parallel port to sense data ready and then reads three BCD digits indicating the intensity of the signal at that point. The LAM then enables the densitometer to take another reading with another bit of the parallel port. This instrument is used to read x-ray diffraction films. In addition to the two LAM's there are two terminals which are used for calculations of lattice parameters, d-spacings, etc., as well as program development.

ELECTRON MICROPROBE

The LAM is interfaced to a Cameca electron microprobe which is used extensively for quantitative analysis. The LAM controls the stage in x and y and any one of four crystal spectrometers using three built-in stepper motor controllers. Data from four pulse height analyzers is collected using four built-in timer scalers.

SURFACE ANALYSIS

Three Physical Electronics Auger electron spectrometers are interfaced to LAM's. The LAM's control the energy analyzer using DAC's and sense a analyzer voltage by reading the BCD out put of a digital volt meter using a parallel port. Data is collected using 12-bit ADC's, voltage to frequency ADC's or pulse counting using timer scalers.

DIFFERENTIAL SCANNING CALORIMETRY

Two types of calorimeters have been interfaced, a DuPont 990 thermal analyzer and a Perkin Elmer DSC-2. For both, 12-bit ADC's are used to collect the data and start and stop are sensed by bits of a parallel port.

INFRARED SPECTROMETRY

Perkin Elmer 521 and 457 infrared spectrometers are interfaced using up/down counters to count shaft encoders which are placed on the amplitude and wavelength drives. The up down counters are read with parallel ports.

MASS SPECTROMETRY

The LAM uses a 14-bit multiplexed ADC to read the analog signal from the electron impact detector, the hall probe voltage and the accelerating voltage of a CEC 21104 mass spectrometer. The LAM finds the peaks on the fly and converts them to mass and intensity at the end of the run.

X-RAY FLUORESCENCE

A Siemens SRS-1 X-ray fluorescence spectrometer has been interfaced using a Siemens logic controller and a LAM. The logic controller receives a string of 18 ASCII numbers defining an operation. At the end of the operation it transmits an

8-digit time and 8-digit count. The LAM is interfaced using a serial port which is RS-232 compatable. The operator can program the LAM to issue a series of commands and buffer the data returned.

TRANSMISSION ELECTRON MICROSCOPY

There is a Tektronix 4010 Graphics terminal in the TEM lab which is used for interactive indexing and simulation of electron diffraction patterns and program development.

SYSTEM INTERCONNECTION

Figure 4 shows a typical application in the LAS system. A terminal is connected to serial port 1 (the "terminal" port) and the LAS host is connected to port 2 (the "host" port). The LAM can communicate with the LAS host through the "host" port or the two ports can be linked together via software so that the terminal appears to be connected directly to the LAS host should the user want to interact with the LAS host. At the beginning of the day, the user enters this mode and logs on to the LAS system with a unique account number and password. This enables the user or the LAM to interact with the LAS host. The various interfaces are implemented using off-the-shelf Multibus compatable modules or specialized on wire wrap boards.

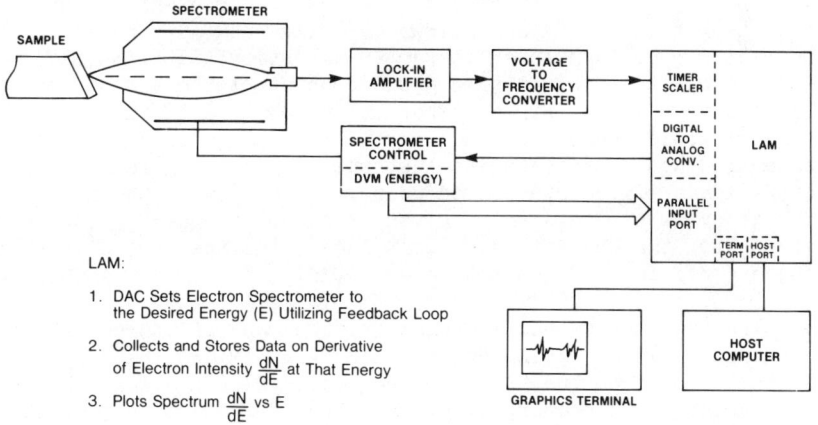

Figure 4.4. Auger Spectrometer connected to LAS

OPERATING SYSTEM SOFTWARE

The LAS host runs under Dynamic OS/32MT, Perkin-Elmer's multitasking operating system. The operating system has such standard features as dynamic memory management, roll in/roll out, overlays and input/output spooling. In addition, there is a Command Substitution System (CSS) which allows the user to create a new system command that is a combination of commands. The Multi-Terminal Monitor (MTM) subsystem allows multiple users and LAM's access to the operating system. MTM allows individual users to operate autonomously, with complete privacy from all other users of the system. In particular, each user can create private disc files, inaccessable to any other user.

APPLICATIONS SOFTWARE

Programs for instrument control and data acquisition which reside in the LAM are written in BASIC with calls to assembly language routines for data collection when necessary. These programs can be down loaded from the LAS host or loaded from the LAM's cassette tape. Likewise data can be transmitted directly to the LAS host or stored on cassette tape. Therefore it is possible for data collection to continue even if the LAS host is down.

A data management system was set up to provide a uniform way of storing data gathered by the microcomputers. Since the operating system software allows a file name with an eight character root and three character extension (FILENAME.EXT), it was decided that a file name would consist of the encoded month, day and year in the first three characters and the spectrum number would make up the next five characters. The file extension would indicate the instrument type (Figure 5). For every data file stored on disc, there is an entry made in a directory file. this entry gives the volume name (the name given to the physical disc cartridge), the spectrum number and space for a forty eight character title.

The data management software is made up of three FORTRAN programs MICRO, DIRSCH and CLEANUP. MICRO handles data transmission from the LAM's to the Host. DIRSCH allows the user to examine the directory file, and CLEANUP removes deleted files from the directory.

A LAM initiates a data transfer by sending the word "MICRO". This causes a CSS file to be executed that loads and starts the

MICRO program. The LAM waits for a prompt (".") from the Host indicating it is ready to receive the file name from the LAM. The program then checks to see if there is already a file by this name on disc. If it does, the operator is notified and the file is deleted and a new one is created with the same name. The directory file is opened and the file is positioned to overwrite the existing entry. If the file name did not previously exist, the program simply creates a file and opens the directory file at the end. The program then sends a prompt to the LAM for the title and running conditions (sample size, flow rate, sensitivity, etc.). The file name and title are written into the directory file and the title and running contitions are written into the data file. The data is then transferred in records with a prompt after each one. When the transfer is complete, the directory file and data files are closed and the micro program goes to end of task. MICRO was written so that new instruments could be easily accommodated. When a new instrument is brought on line, the only changes needed are in the decoding of the instrument type and reading and writing of the running conditions.

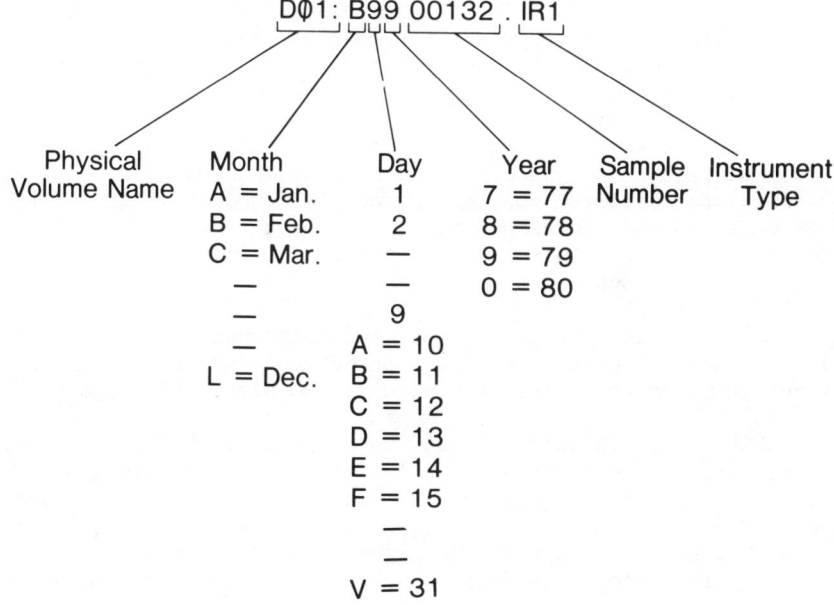

Figure 4.5. Convention used for naming date files.

DIRSCH allows the user to search the directory file for information about the data stored on disc. The search is done by means of one letter commands. The "H" command returns the highest sample number currently on disc. This is useful in assigning new sample numbers. All of the other commands return the entire directory line, (i.e. file name, month, day, year, and title). The month, day and year are not actually stored in the directory, but are decoded from the file name. The "L" command lists the entire directory. "D" lists all entries on a specified date (D 12/28/76). The "S" command lists all entries with a designated sample number (S 12345). The "I" command lists all entries for a given instrument type (I LC1). The "K" command deletes a given entry from the directory. This command actually places a question mark "?" as the first character in that line causing this entry to be ignored.

The H, D, S, I and K commands deal with the file name that occupies the first sixteen characters of the file. The other forty-eight characters are the title, which includes any designation chosen by the user to identify the particular sample. The "T" command is the title search command. It can be used to search the title for a particular string. For example "T:PITT OIL:" will cause all the entries containing "PITT OIL" anywhere in the title to be printed out. The "E" command terminates the program.

DIRSCH is written in modular form such that each command is processed by its own subroutine. These subroutines can be overlaid to achieve the best economy of core storage. If the user wants to add another command, it then becomes a simple process of adding this command to the main program and writing the appropriate subroutine.

CLEANUP is used to "clean up" and condense the directory file. This program rewrites the directory line by line, excluding lines that were deleted with the dirsch "K" command.

SPECPLOT is a generalized data reduction and plotting program which allows the user to interact directly with spectral data. The program is designed to handle a wide variety of data types (Figure 6). The data type (DIF,DSC,IR1,LC1,etc.) is stored along with title and running conditions in the first record of the file. The type designation is used to put the correct axis labels on plots, select correct scaling factors, select correct program options and any information required from the user for the particular type of data being manipulated. The program is written for use with a Tektronix 4010 Graphics terminal which has X and Y cursor controls and 4631 hard copy unit. Optionally, the program will support a Tektronix 4662 plotter in series with the terminal so that either can be used.

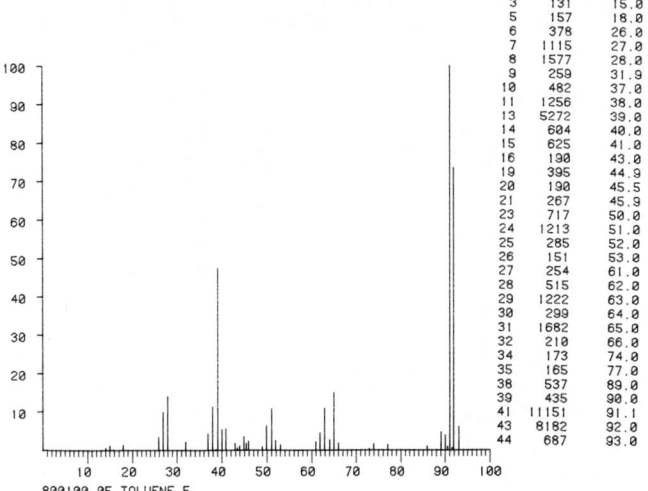

Figure 4.6.a.   Spectrum from Mass Spectrometer.

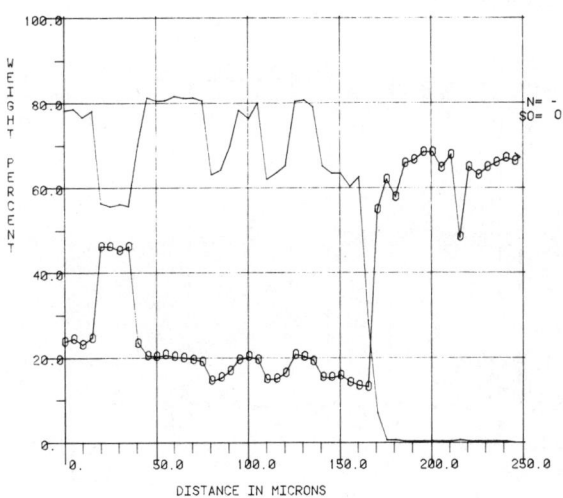

Figure 4.6.b.   Electron Microprobe diffusion profile.

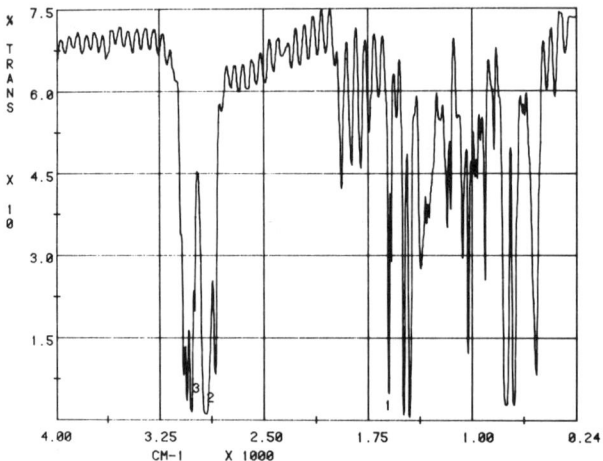

Figure 4.6.c. Infra Red Spectrum.

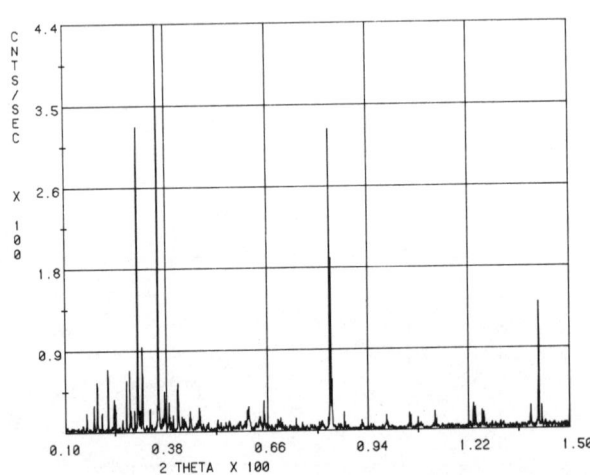

Figure 4.6.d. Powder Diffractometer Scan.

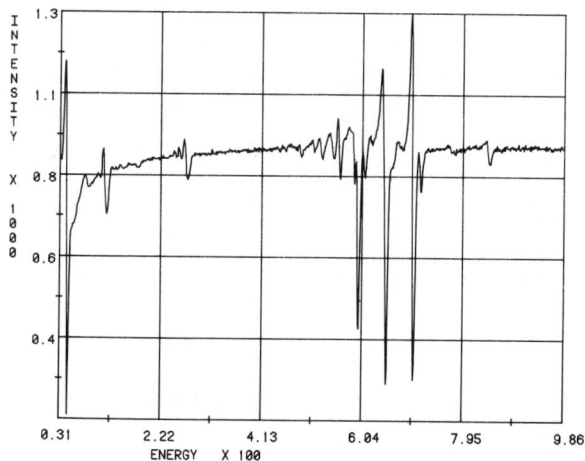

Figure 4.6.e.   Auger Electron Spectrometer Scan.

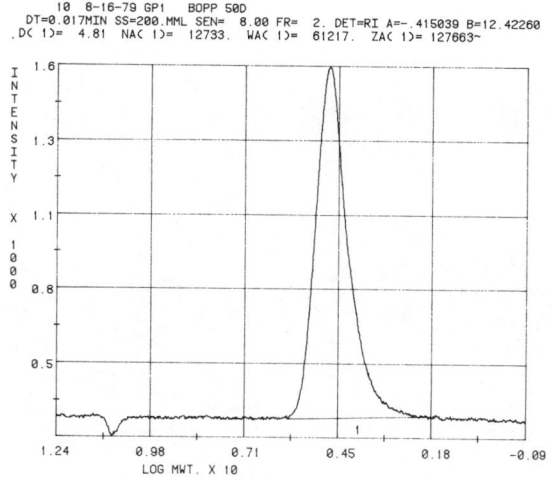

Figure 4.6.f.   Gel Permeation Chromatogram.

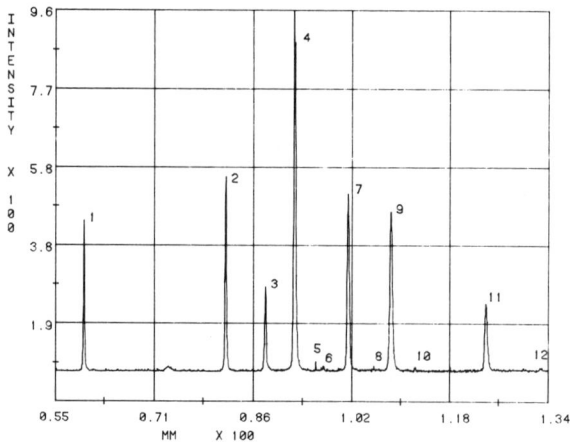

Figure 4.6.g.   Densitometer Scan.

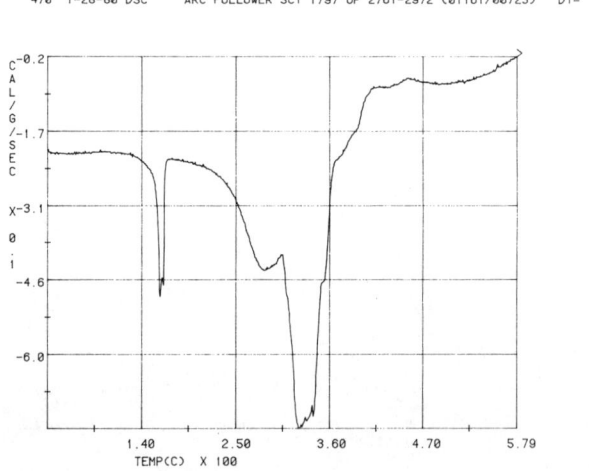

Figure 4.6.h.   Differential Scanning Calorimeter Scan.

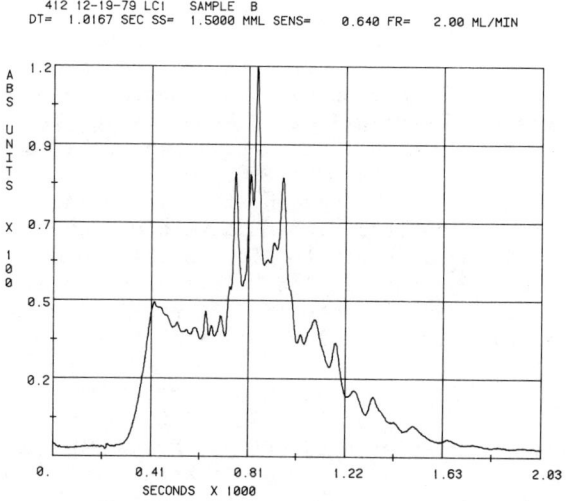

Figure 4.6.i. Liquid Chromatogram.

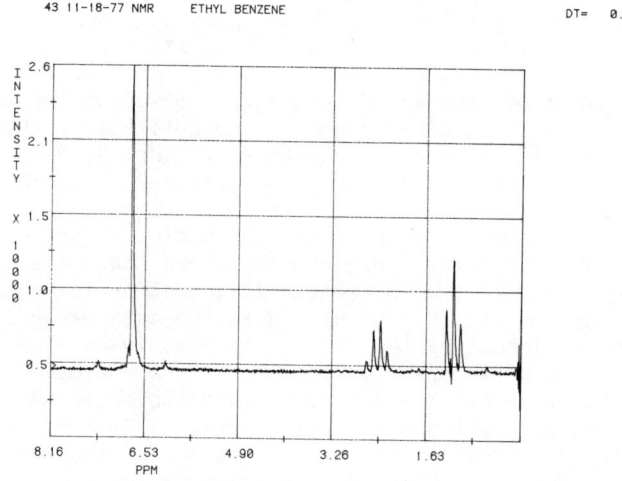

Figure 4.6.j. Nuclear Magnetic Resonance Spectrum.

SPECPLOT is written in modular form. Presently it consists of a root program and twenty eight subroutines. This was done to make additions easy and to allow the program to be extensively overlaid if necessary. The resolution on the screen of the 4010 graphics terminal is 1024 points in the X-direction and 780 in the Y-direction. When plotted full screen, the actual spectrum is 936 by 630. The rest of the space is used for labeling, sample identification, title, running conditions, and results returned from interactive commands. The program buffers up to 4096 data points to give quick access for data reduction, display and interaction. This value is arbitrary and can be easily increased or decreased. A provision has been made to handle more data, but at a loss of resolution.

The program displays data in three modes, full screen, split screen or overlaid. The overlay will plot the two spectra with the second spectrum displaced slightly. No y-axis is provided as it has no meaning. For both split screen and overlays, the first spectrum given is the master and the second is the slave. Thus the two spectra are plotted to the scale of the first data file (Figure 7). The addition option provides a means of plotting composite spectra while the subtraction option provides a means of removing one component from a spectrum (Figure 8).

A linear equation has been provided for correcting the x-axis. the form of this correction is $A*X + B + X$. It allows the user to correct for drift that may occur in the instrument over the course of an experiment. SPECPLOT provides some totally automatic data processing capabilities. These include least square moving average smoothing (Figure 9) (2), background subtraction (Figure 10) (3), K-alpha-2 elimination for diffraction data (4), a re-zero to first peak for NMR and chromatographic data and a second derivative search algoritham (Figure 11) (5,6,7). Figure 12 shows the dialog used in selecting these various options. If an option is not desired, a carriage return is entered. When the data reduction options have been performed by the program, a plot is made of the data as it appears after these options have been executed. Following the plot, moveable X & Y cursor lines appear on the screen and the program enters its interactive mode. A list of the commands which have been written to process the data in this mode is shown in Table 2. Figure 13 shows examples of the expand and integrate options. If the search option was used, a tabulation of the results of the search appears on the screen and is written to a data file for later printing if the user desires.

In addition, the peaks of the plotted spectrum are numbered corresponding to the peaks found in the search (Figure 11).

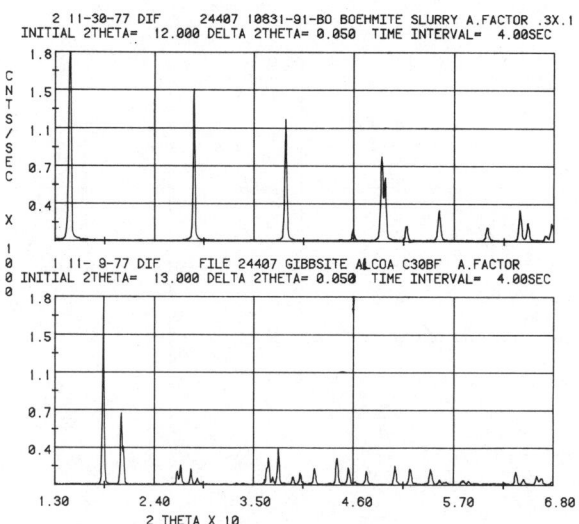

Figure 4.7.a. Example of Split Screen Display Mode.

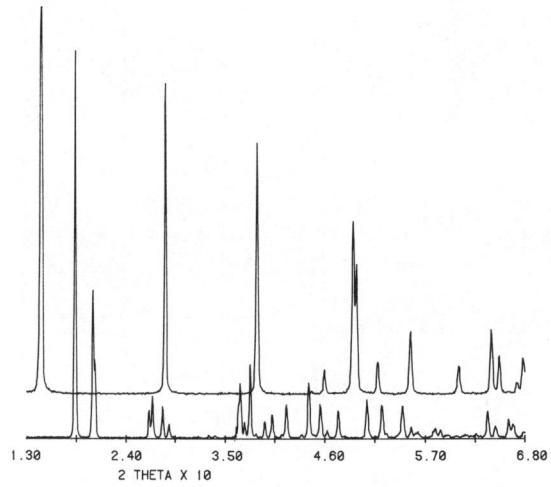

Figure 4.7.b. Example of Overlap Display Mode.

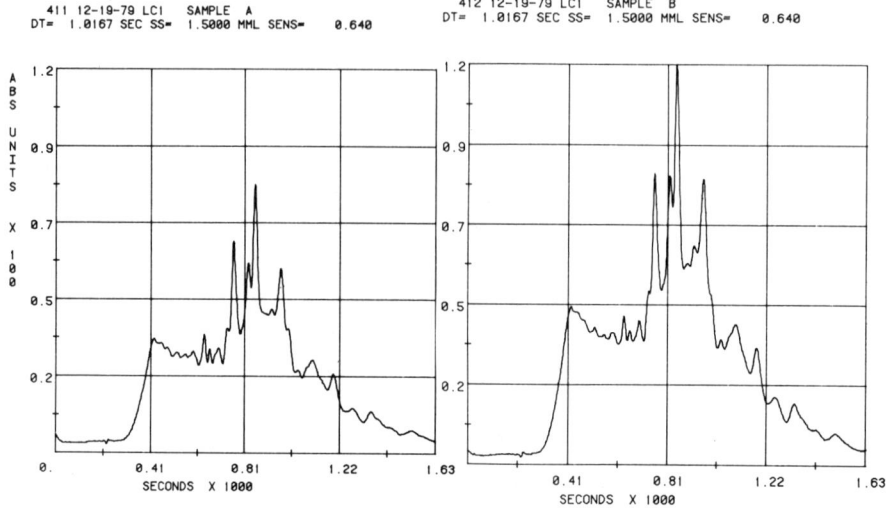

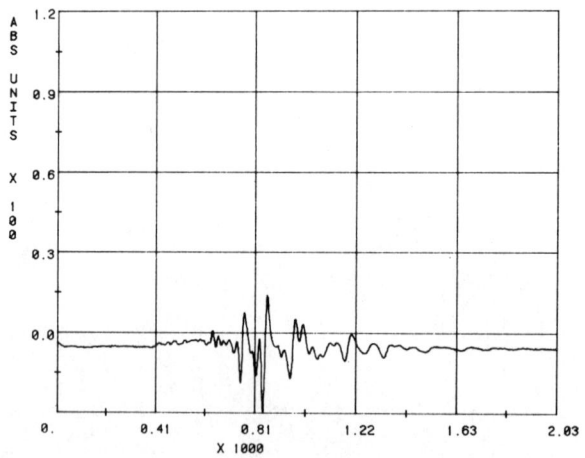

Figure 4.8. Subtraction of Two Spectra.

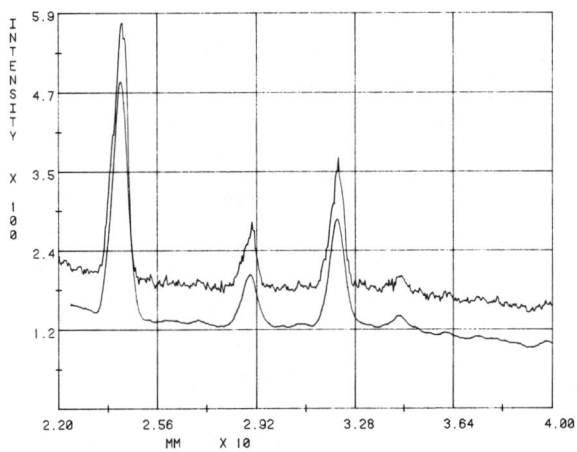

Figure 4.9. 25 Point smooth of data.

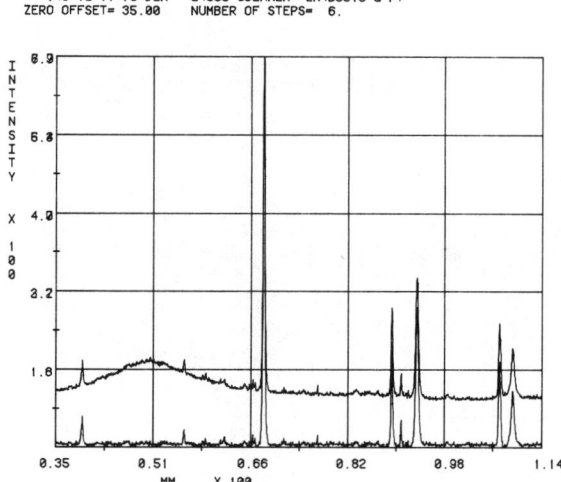

Figure 4.10. Background Subtraction.

N	POSITION	HEIGHT	AREA	P	W2	W4
1	59.628	368.000	96.432	271.998	0.317	0.262
2	82.150	472.000	149.688	392.623	0.352	0.317
3	83.390	200.000	64.218	164.205	0.352	0.321
4	93.008	879.000	370.902	839.924	0.458	0.422
5	96.469	20.000	3.318	14.527	0.317	0.166
6	97.669	10.000	2.142	5.534	0.282	0.214
7	101.641	428.000	171.360	390.218	0.423	0.400
8	105.775	8.000	1.092	4.751	0.352	0.136
9	108.500	387.000	197.442	401.686	0.528	0.510
10	112.375	7.000	1.638	4.726	0.282	0.234
11	123.775	161.000	80.094	152.622	0.493	0.497
12	132.737	6.000	1.806	7.851	0.599	0.301

FOJ
MEAN Y= 0. +OR-0.213E+00 MEAN D2Y= 0.0025 +OR-0.139E-01

Figure 4.11.a. Ouput from Automatic Peak Search.

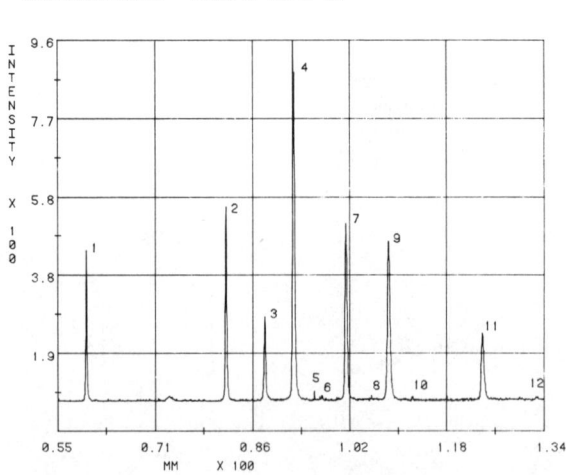

Figure 4.11.b. Plot of Diffractometer data showing peaks labeled according to output above.

```
*RUN SPECPLOT,S                             YMIN=        0.  YMAX=     20112.
06/02/80  14:51:50                          YMIN=
SPLIT SCREEN-1 OVERLAY-9 SUBTRACT-8 ADD-7   >
>                                           YMAX=
DATA FILE                                   >
>D01:JQ800001.NMR                           PLOTTER ?
    1 10-26-78 NMR                          >
UNKNOWN OIL SAMPLE

DELTA X= 0.556E-02PPM
LINEAR CORRECTION
A*X+B+X
A=
>
B=
>
NUMBER OF POINTS= 1500

SMOOTH-00,05,07,09,11,15,17,25
>
ZERO TO 1ST PEAK-1
>
BKGR-1
>
SEARCH-1
>
XMIN=   0.    XMAX=   8.328
XMIN=
>
XMAX=
>
```

Figure 4.12. Typical User Dialog for SPECPLOT.

Table 4.2. Interactive cursor commands for specplot.

T -- Finds the X & Y coordinate for intersection of 3 lines (used for DSC spectra).

X -- Actual X pos'n of vertical cursor.

Y -- Actual Y pos'n of horizontal cursor.

M -- Actual X & Y pos'n

I -- Two cursor pos'ns specify the range of integration and background. This command returns the X pos'n of the highest point, the Y height above background of this point and the integral of the peak.

A -- The same as I, but it also calculates the Debye half width and centroid of the peak.

C -- the same as A, except the actual Y value of the curve at the 2 cursor inputs are used for the background.

Z -- Integrates between 2 cursor pos'ns everything above zero and returns only the integral.

W -- Weight, number and Z average molecular weight. Input same as I (used for GPC analysis)

F -- Fits peak to a gaussian & returns its pos'n, integral, height and half width. 4 cursor pos'ns are required, 2 for the background and 2 for the range of fit.

E -- Expands the plot specified by 2 cursor inputs.

S -- Shifts the plot left by 2048 points if the plot contains more than 4096 points.

O -- Actual X pos'n of the cursor cross hairs.

P -- Parabolic fit to the top of the peak.

R -- Replots the data between XMIN, XMAX and YMIN, YMAX.

G -- X pos'n in angstroms for Cu K-alpha 1 radiation (used for diffractometer).

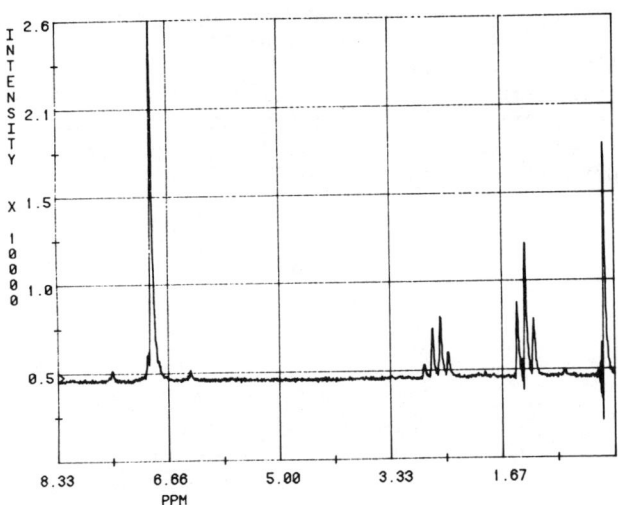

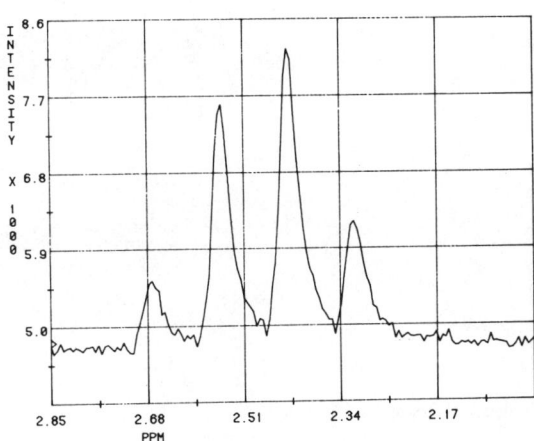

Figure 4.13. The lower plot is an expansion of the quartet in the upper plot. The first three peaks were integrated using the interactive cursor options.

98   The GE Laboratory Automation System

In addition to programs directly related to storing and processing of data collected by LAM's, the system is also used for other computational needs. PATTERN and KIKUCHI are two programs used extensively by the Transmission Electron Microscopy (TEM) group. PATTERN is used for the indexing and simulation of electron diffraction patterns (8,9,10). KIKUCHI is used to simulate stereographic projections. LAUE is used by the X-ray diffraction group to simulate x-ray diffraction patterns (Figure 14) (11). Sophisticated molecular orbital calculations are run in background during the day and in foreground during off hours (Figure 15) (12,13).

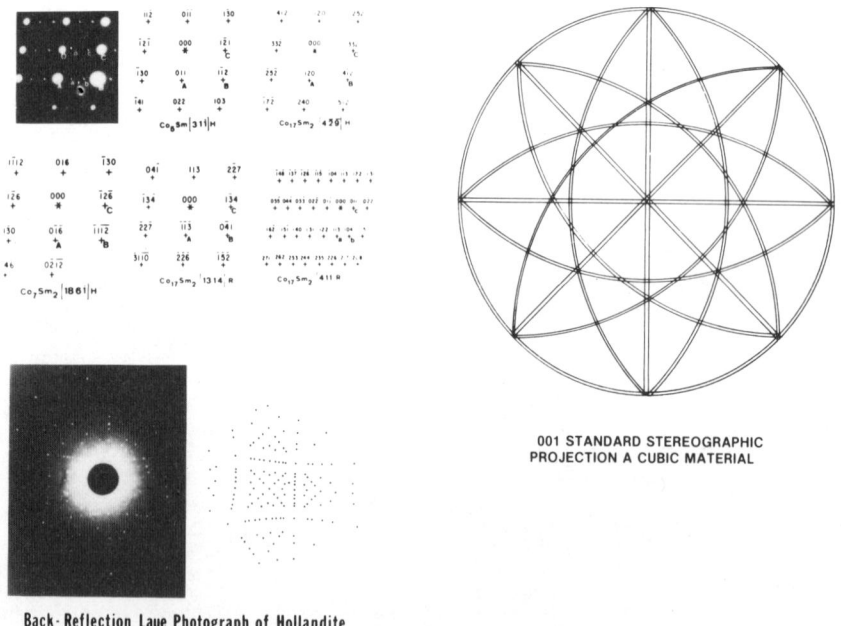

Figure 4.14. Computer Generated Diffraction patterns.

Applications Software 99

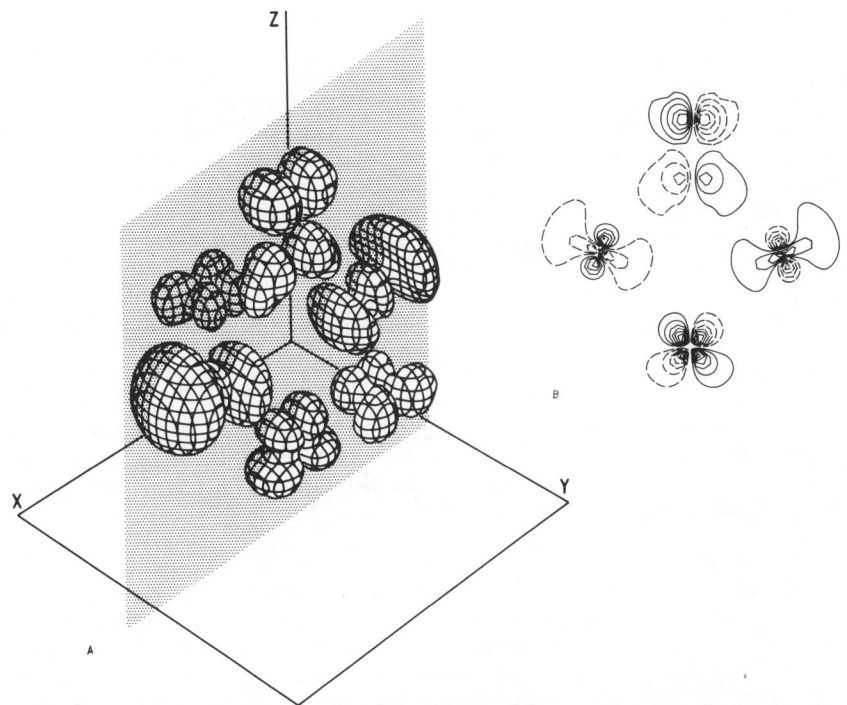

Figure 4.15. Contour plots of the highest occupied molecular orbital for CU5CO cluster. The CU5CO cluster is a model for studying chemisorption of CO on (100) face of CU metal.
A.) 3D-plot in perspective.
B.) 2d-plot of orbital in XZ-plane.

In addition to FORTRAN compiler, BASIC interpreter and editor, the host has a powerful word processing program called TEXT. TEXT is used by the various people in the branch for writing reports, memos to customers, etc. It is especially useful when people are collaborating in writing a report. In fact, this chapter was written using the text processor.

CONCLUSION

The Laboratory Automation System here was developed at a hardware cost of approximately two hundred thousand dollars. The initial software development was about a three man year effort however the software is continually being updated and new programs are always being added. A LAM required to add a new instrument would cost between three and five thousand dollars depending on the application. A "turn key" data system for an instrument can easily cost between twenty and thirty thousand ollars and will generally only function with one instrument or type of instrument. One could easily invest two hundred thousand dollars in separate data systems and still have no guarantee that these systems can communicate with each other or with any other computing system. The LAS provides data acquisition and processing for a wide variety of analytical instruments and new instruments can be added with relative ease and at a small incremental cost. In addition the system has raw computing power of large timesharing systems. In our lab, we have been able to move nearly all of our activities previously run on timesharing to the LAS host.

The system is designed to make it possible to easily add new instruments. We recently purchased a new Cameca microprobe with a DEC based "turn key" data system. Because of the design of the LAS, it was quite easy to interface this data system to take advantage of more sophisticated data reduction programs available in the LAS host. Future plans call for the addition of gas chromatographs in our separations lab and our Fourier Transform NMR's. In addition we will be adding search capability in our diffraction lab using the JCPDS file.

Final Note:

The camera-ready copy for this chapter was generated by the text processor on our LAS host and transmitted to a Xerox 850 typewriter in our branch office.

REFERENCES

1. W.T. Hatfield, M.F. Ciccarelli, R.B. Bolon and E. Lifshin, "A Real Time Approach to Laboratory Automation," Proc. Ninth Annual Conference of the Microbeam Analysis Society (1974).

2. A. Savitzky and M.J. Golay, Analytical Chemistry 36, 1627-1639 (1964).

3. R.P. Goehner, Analytical Chemistry 50, 1223-1225 (1978).

4. W.A. Rachinger, Journal of Scientific Instruments 25, 254-255 (1948).

5. M.A. Mariscoth, Nuclear Instruments and Methods, 50, 309-320 (1967).

6. S.J. Mills Nuclear instruments and Methods 81, 217-219 (1970).

7. E.J. Sonneveld and J.W. Visser, Journal of Applied Crystallography 8, 1-7 (1974).

8. R.P. Goehner, W.T. Hatfield and P. Rao, Proceedings of the 34th Electron Microscopy Society of America, 542-543 (1976).

9. R.P. Goehner and P. Rao, Metallography 10, 415-424 (1974).

10. P. Rao and R.P. Goehner, Journal of Applied Crystallography 17, 482-488 (1974).

11. R.P. Goehner, Advances in X-ray Analysis 19, 725-733 (1976).

12. R.P. Messmer and S.H. Lamson, Chemistry and Physics Letters 65, 465 (1979).

13. R.P. Messmer, S.H. Lamson and D.R. Salahub, Has been accepted for publication in Solid State Commun.

5

WARPATH, A COMPUTER NETWORK FOR A MULTIUSER LABORATORY ENVIRONMENT

George C. Levy, Dan Terpstra, Charles L. Dumoulin

INTRODUCTION

Laboratory computer networks may be established to serve various needs including:
1. Control of experiments;
2. Data acquisition;
3. Data storage and management;
4. Data reduction with subsequent outputs to the user;
5. Other tasks such as laboratory communications, reports generation, batch processing for number crunching and other uses.

The design of a computer network is predicated on the particular mix of these tasks <u>and on the required throughput</u> for these operations. Thus, a network designed for data logging slow chromatographic data from a small number of experimental stations will have very different requirements from a network set up to mix a large number of fast and slow data acquisitions requiring manipulations and transfers of many large data sets. Other constraints may, of course, also affect the design of computer networks, including the overall physical size of the configuration, the number of required nodes, the total number of network transactions per unit time, etc.

The WARPATH local area computer network was designed to maximize efficiency under the following conditions:
1. Accommodation of up to 15 nodes is required, with flexible spatial arrangement (total network length 1 or more kilometers).
2. Data transfers can be quite large (e.g., up to 64K bytes per data file), interspersed with other network transactions consisting of short (<1K byte to 4K bytes) messages, initializations, etc.
3. The rate of network access is only moderate (integrated to less than 5 transactions per minute but traffic can be as heavy as several transfers per second).

4. Laboratory data acquisition is assigned to inexpensive microcomputer stations which have capabilities for fast dual channel data accumulation (acquisition/averaging). For economic reasons, these stations do not have local mass memory, but rely instead on network shared-resources for storage of large data sets.
5. All higher level data reduction/data processing is assigned to a large minicomputer running a multi-user, multiprogramming operating system. This computer, which is one of the network nodes, itself utilizes distributed processing, with intelligent soft and hard output devices.
6. Shared resource mass data storage for the network resides in both the large minicomputer and in one microcomputer configured as a development system/network hub but distinct from the specific microcomputer controlling the network.

This chapter summarizes the overall hardware and software characteristics of the WARPATH data network. Current developments in local area networking are occurring so rapidly that any existing implementation is inherently obsolete. Nevertheless, it is hoped that the discussion below will serve as one stepping stone for the design of future laboratory computer networks.

THE WARPATH LOCAL AREA NETWORK

General Considerations. An effective means of transferring data between multiple computer systems is essential to the development of a distributed resource facility such as that being implemented in the FSU nmr laboratory. To that end, the WARPATH Local Area Data Network was developed. Although the acronym is somewhat tongue-in-cheek, it serves to illustrate some important aspects of this network configuration. These aspects of the WARPATH (for: Widespread Asynchronous Rapid Accurate Transfer Handler) are presented below:

WIDESPREAD - Differential line transceivers are used throughout the WARPATH, allowing a total network length of up to 4000 meters distributed among up to 15 distinct nodes.

ASYNCHRONOUS - The WARPATH employs a derivative of Hewlett-Packard's patented 3-wire handshake, insuring a fully interlocked transfer of information between computers operating at different speeds. A flow chart of this handshake is illustrated in Figure 5.1.

RAPID - Powerful Z-80 block I/O instructions are utilized at every node, allowing the WARPATH to handle data at burst rates of 190K bytes/second. Overall data throughput approaches 50K

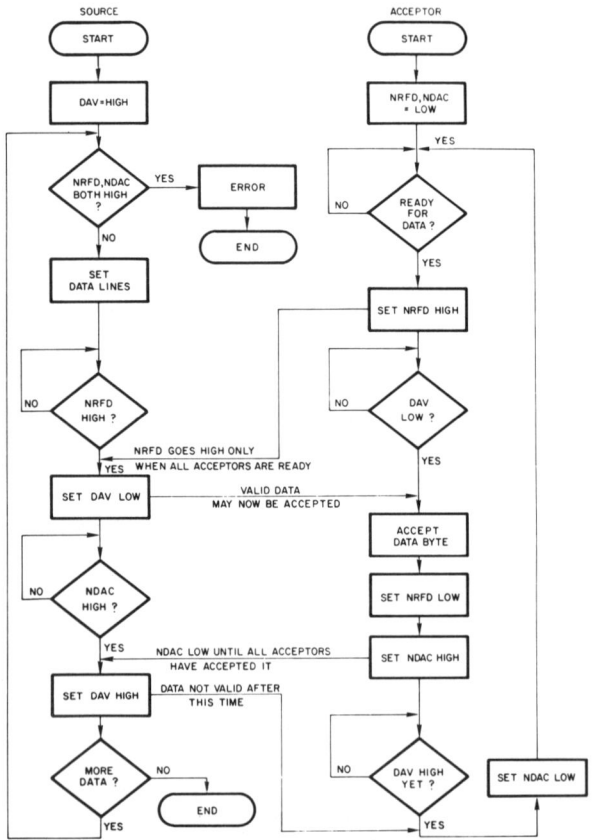

FIGURE 5.1.   Flow chart for the GPIB three-wire handshake employed on the WARPATH data network. One source node can transmit to any number of acceptor nodes with no loss of data.

bytes/second. The graph in Figure 5.2 illustrates this data rate as compared to other common data transmission rates.

PARALLEL - A bit-parallel, byte-serial transfer structure sends data on the WARPATH one byte at a time, greatly enhancing both data integrity and transfer rates as compared to traditional bit-serial techniques.

ACCURATE - Differential transceivers provide effective common-mode noise rejection resulting in error-free data transfers. Over 10^{10} bytes of data have been transmitted in error-checking tests with no observed errors. In addition, the WARPATH has been in limited operation for almost a year with no

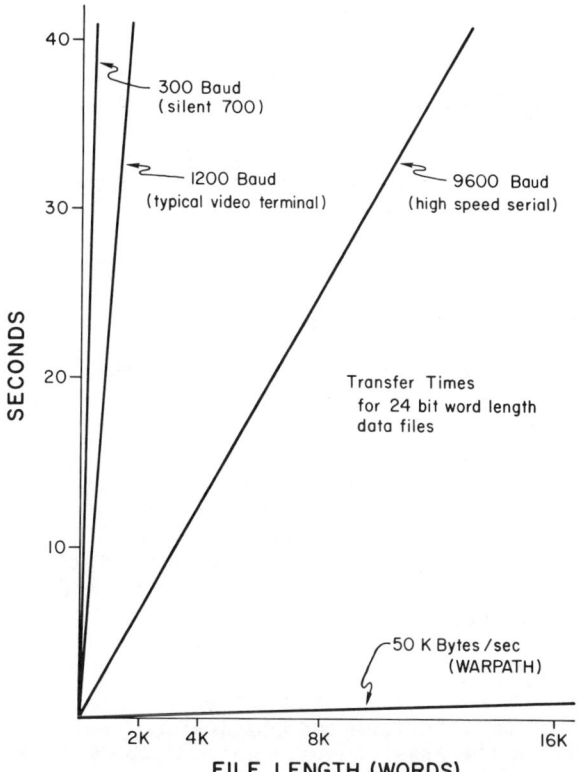

FIGURE 5.2.   Transmission time versus file length for the WARPATH data network as compared to typical serial (RS-232) transmission rates.

data transfer errors reported. In the remote case of an error, detection and recovery occurs through use of 8-bit checksums that are sent with every network transaction.
  TRANSFER HANDLER - Basic transfer protocols (link layer) are implemented by the Intel 8291 and 8292 Large Scale Integrated (LSI) GPIB Interface chips. Application-specific protocols (network and transfer layer) are handled at all nodes by Z-80A microprocessors, making staightforward any device-dependent interfaces. This is discussed in further detail in the section on network protocols.
  The introduction in 1972 of Hewlett-Packard's General Purpose Interface Bus (GPIB) led to an early international

standardization of the protocols for this bus as the IEEE-488, adopted in 1974. This standardization led in turn to the integration of GPIB interface logic into LSI circuitry around 1978. These LSI chip sets are now available in various forms from Texas Instruments, Motorola, Intel, and others.

The standard GPIB implementation, with its distance limit of 20 meters, is too restrictive to be employed as a true local area data network. Bus extender devices can be purchased from various sources, but they are generally expensive and topologically restrictive. They often also result in drastic reductions of data throughput rates. Single chip GPIB interfaces make it feasible for the designer to employ concepts and protocols of the GPIB as the heart of a new local network, incorporated into a transmission medium without the distance restrictions of the standard GPIB structure. This is the approach that has been taken in the development of the WARPATH network. Such an approach inevitably implies hardware specialization, and some design compromises or sacrifices must be made. This is unavoidable, since no commercial product available at the time this project was undertaken fulfilled the necessary requirements of speed, distance and flexibility. Within a few years, however, high-speed serial network protocols such as Xerox's Ethernet will have become sufficiently established to warrant LSI circuitry and simple implementation of local networks with none of these restrictions and compromises.

WARPATH: The Physical Layer. The physical layer of any network involves the actual transmission medium and signals needed to send information from one physical node of the network to another. This is the layer to which attention must be paid in converting the GPIB into a useful local network configuration.

Typical GPIB signal generating circuitry is demonstrated in Figure 5.3. The bus driver is enabled by the READ/WRITE signal as shown. The open-collector output of this driver implies that a low logic level imposed by any bus driver overrides a high logic level asserted by any or all other drivers. This "wire-and" function is essential to the overall operation of the GPIB data transfer sequence. The circuitry used to implement the bus interface, and the termination resistors shown in Figure 5.3, lead directly to the standard GPIB limit of 15 devices and 20 meters total length.

In order to overcome the length restrictions of the standard GPIB, the WARPATH utilizes differential transceivers identical to those specified for use in the new high-speed serial standard, RS-422. These transceivers offer long distance transmission characteristics of several thousand meters, as well as the high noise immunity required in a noisy environment such as an nmr laboratory. The major drawback of differential transceivers in

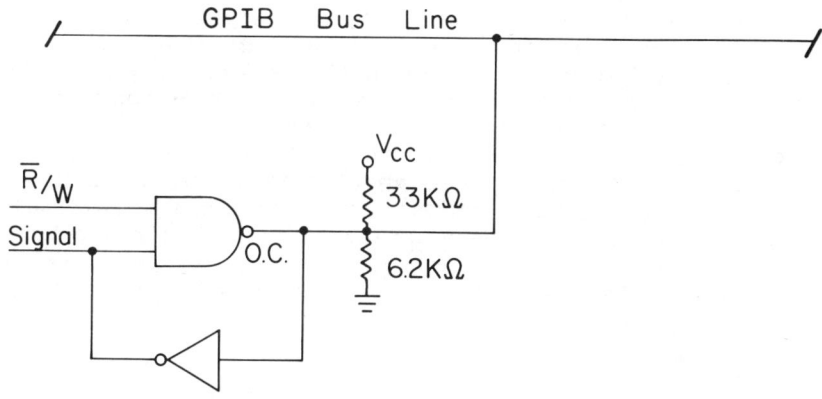

FIGURE 5.3. Standard GPIB bus interface. The NAND gate is an open-collector device, implementing the "wire-and" function, as discussed in the text.

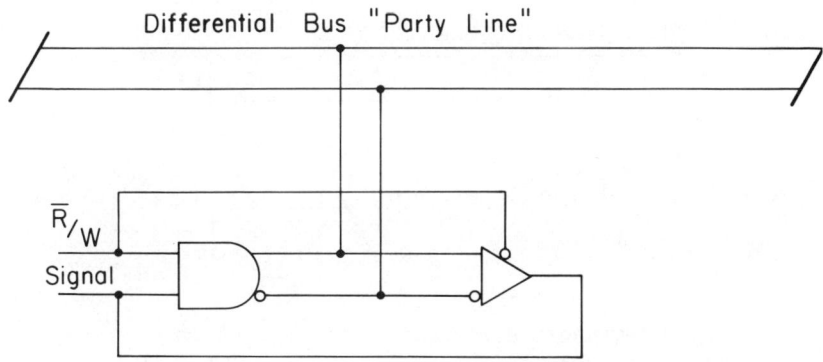

FIGURE 5.4. WARPATH differential party-line interface. Only one node can drive the party-line at any time, and "wire-and" capability is not required.

the context of the GPIB lies in the fact that they cannot directly perform the <u>wire-and</u> function described above. Thus, two types of transmission circuitry are required for the WARPATH. The majority of signals, including the eight data lines and various control lines, are implemented with standard differential party-line techniques as shown in Figure 5.4. This party-line forms a continuous bus connecting all of the nodes of the network, with line-termination (to supress signal reflections) at either end of the line. The remaining three signals require the <u>wire-and</u> function and are implemented as shown in Figure 5.5. The differential signals from each active node are transmitted to a common hub, where they are converted into electronic levels that can be <u>wire-and</u>ed. The result of this <u>wire-and</u> is then retransmitted to every node to provide the differential equivalent of the standard GPIB open-collector signals.

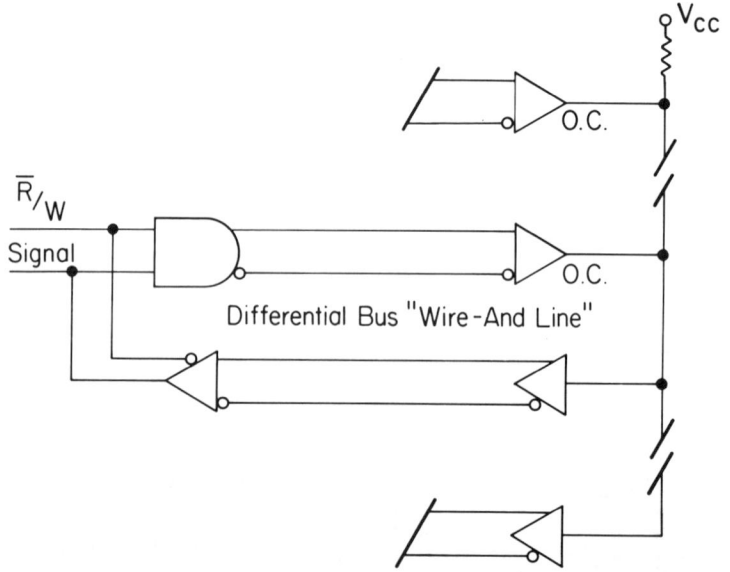

FIGURE 5.5.   WARPATH differential "wire-and" interface. Since differential signals can not be wire-anded directly, the 3 bus signals that require this feature are converted to single-ended signals, wire-anded, and reconverted to differential signals enabling the full features of the three-wire handshake shown in Figure 5.1.

The WARPATH Local Area Network

<u>WARPATH</u> <u>Topology</u>.   The topology of the WARPATH data network does not fall easily into any single traditional topological category. Both physically and logically, the topology shares features of "bus" or "broadcast" networks and features of "star" networks.

In terms of physical layout, the two signal types mentioned earlier (party-line and <u>wire-and</u>) follow independent paths. The party-line signals make point-to-point connection with each node, forming a bus topology. The <u>wire-and</u> signals must meet at a common point before being redistributed, in a configuration similar to a star, except that the point of intersection need not be an actual node of the network. This leads to a conceptual distribution in which pairs of network nodes form the base of a triangle with the central termination point at its apex. Several nodes in the network then lead to a roughly "propeller" shaped physical network topology as shown in Figure 5.6.

The logic of the GPIB bus is retained in the operation of the WARPATH, resulting in a network in which a single "controller" node serves as the logical hub of a star, controlling all communication that occurs on the bus. However, the fact that all

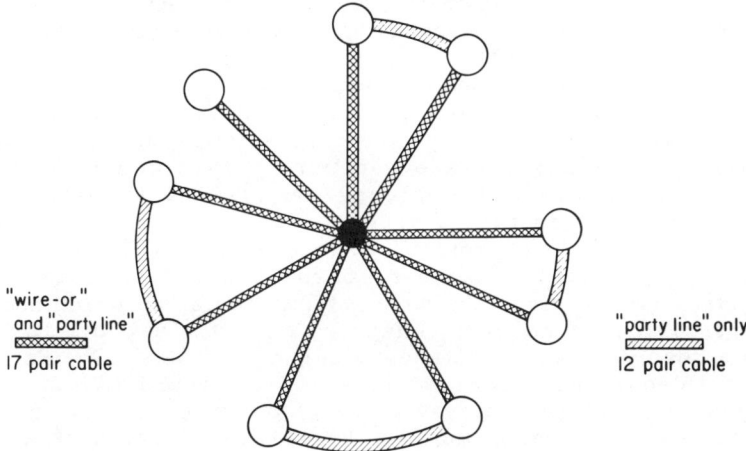

FIGURE 5.6.   The "Propeller" configuration of the WARPATH data network. All physical wiring meets at a common point, shown by the black circle, where the necessary signals are wire-anded. Party-line signals are not connected at this point, but form a multi-drop bus connecting all nodes of the network.

nodes of the WARPATH are interconnected through the bus medium implies that messages are broadcast to every device, and received only by those that have been addressed by the controller. This differs from a typical star network topology in which all network messages must pass through the central node.

The WARPATH topology has the inherent disadvantage of requiring a controlling device. This implies that network operation is linked to the functionality of that device, and a failure at that node disrupts all network communication. In this case, as is seen below, the central controller consists of minimal hardware, simplifying repair and making back-up duplication inexpensive. The central controller also proves to be advantageous in that communications protocols and resolution of bus conflicts become much simpler when handled by only one device in the network.

The WARPATH CHIEF. One node of the WARPATH network serves a function substantially different from any other. This node is called the Communications Host and Information Exchange Facility (or WARPATH CHIEF). This node serves to establish the data links (often called "virtual circuits") between communicating devices, ensure that network transactions are carried out in an orderly manner, recover from network error conditions when possible, and inform users of the status of various network functions. The CHIEF is completely autonomous in normal operation, requiring no user interaction at all.

A minimal CHIEF system consists of two S-100 microcomputer circuit cards: the network Controller Talker / Listener (CT/L) card and a single-card Z-80 based microcomputer containing RAM, ROM, and serial interface hardware. This computer card is identical to the one employed in the INDIAN data station described later in this chapter. Addition of a Video Graphics card and a television monitor to this minimal CHIEF enhances its utility by providing access to the built in status display software, which gives the user information as to which nodes are currently in the network, whether or not the network is in use, and a record of the most recent network errors. The format of this video network monitor is shown in Figure 5.7. If a video display is employed, a systems level operator can gain access to a machine level debug monitor by simple addition of a serial (keyboard) input to the single-card computer. Such a monitor can be vital in detecting and correcting network errors not automatically handled by CHIEF software. A block diagram for a fully implemented CHIEF is shown in Figure 5.8.

The hardware of the CT/L circuit card consists primarily of the Intel 8291 Talker/Listener chip and the 8292 Controller chip. Additional circuitry implements the differential tansceivers necessary for the physical WARPATH interface and some simple

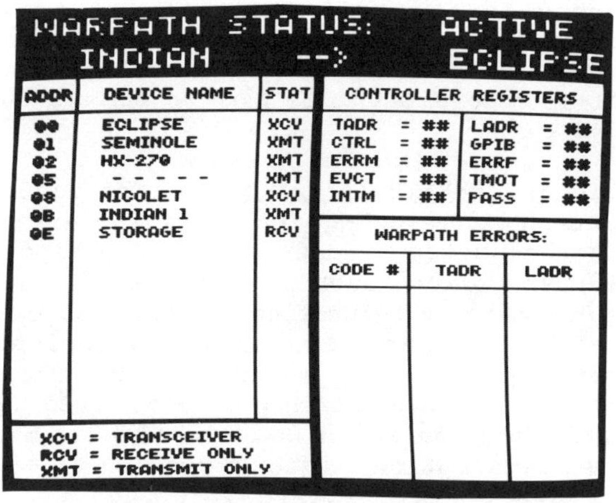

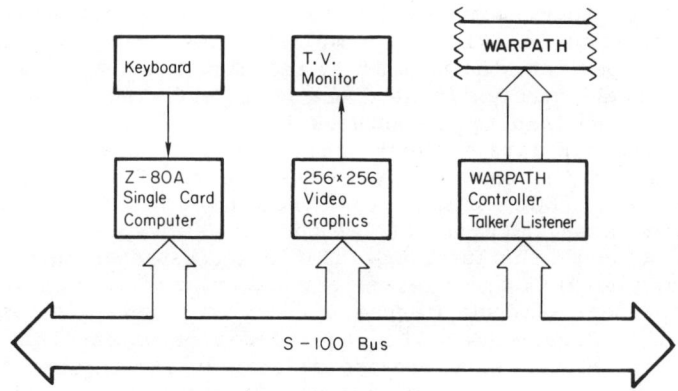

*FIGURE 5.7.   Photograph of the WARPATH CHIEF status display.
The right side gives information on currently
active nodes; the left gives diagnostic information and a history of recent network error conditions.*

*FIGURE 5.8.   Block diagram of the WARPATH CHIEF. The optional
video graphics circuitry provides network status
information and the keyboard input allows access
to network diagnostic programs.*

logic is required to provide the S-100 interface. The intelligence of the Intel LSI circuits reduces external circuitry to a minimum, allowing almost all of the interactions between the CT/L and the microcomputer to be handled through 8-bit input and output ports.

The Intelligent WARPATH Interface. The equipment in many multi-instrument laboratories, including the nmr laboratory at FSU, is often acquired over an extended period of time from several sources. Given such an environment, any attempt at providing comprehensive networking capabilities must address the necessity of interfacing the network to a variety of computer architectures. Specifically, the FSU nmr laboratory includes two 20-bit Nicolet 1080 minicomputers, a 16-bit Data General ECLIPSE minicomputer, and a variety of S-100 based 8-bit microcomputers.

One possible solution to such a problem is the design of a specific hardware and software interface for each type of computer in the network. This solution is inefficient in the case of several computer types in that it requires a major design effort with every addition of a new type of computer to the network. A far superior solution isolates all of the computer-independent network hardware and intelligence into a modular "black box" with relatively simple computer interface requirements. This approach offers two important advantages over the one-at-a-time method: Network interfaces become identical, reducing protocol complexity and increasing maintainability. Also, interface modularity simplifies the incorporation of new computer types into the network by reducing the design problem to the relatively straightforward one of interfacing to the network module, rather than to the network itself.

The WARPATH data network capitalizes on recent advances in microelectronics technology in the design of its intelligent interfaces. Each interface incorporates its own Z-80A microprocessor, 1K bytes of RAM buffer memory, 2K bytes of ROM program memory, the Intel 8291 LSI Talker/Listener chip, and the necessary network transceiver circuitry, making it a complete microcomputer. A block diagram of such an intelligent interface is shown in Figure 5.9. The major function of the intelligent WARPATH interface is to provide a "smart" link between the host computer on one side and the network on the other. A discussion of this circuitry can thus be divided into two areas: the first dealing with the host side of the interface, and the second concerning the network interfacing aspects.

Host: The hardware connections between an intelligent WARPATH interface and its host computer consist minimally of 9 inputs from, and 10 outputs to the host computer. Of these lines, eight in each direction form a byte-wide parallel I/O port

The WARPATH Local Area Network 113

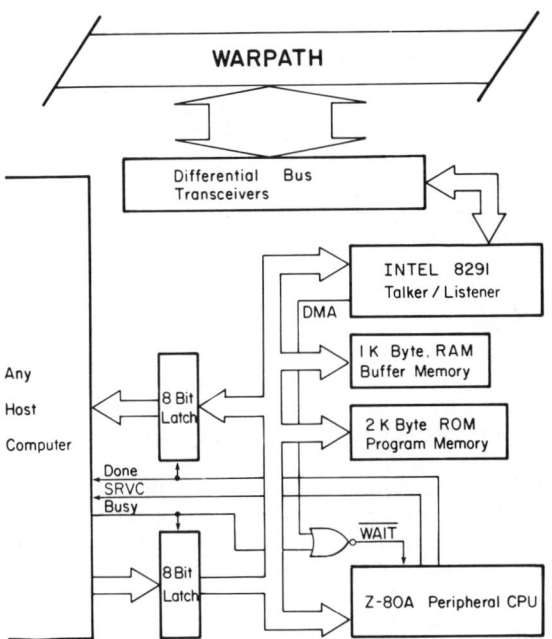

WARPATH Intelligent Interface

FIGURE 5.9. Block diagram of the intelligent WARPATH interface. Host computer interfacing is done through a single 8-bit bidirectional latch and 3 additional control lines. The interface contains its own Z-80 microprocessor, program and buffer memory, and bus interface logic to simplify the demands on the host computer.

through which all data and commands are passed. One line in each direction is used for a BUSY/DONE handshake pair, and the final output line serves as a special service request signal to the host computer. Signals of this sort are readily available on almost any mini- or micro-computer, making this intelligent interface virtually universal. Specific applications may require extra address decoding or other logic to select one of several computer I/O ports for use with the interface circuitry.

The I/O port and handshake lines provide a simple pathway for information exchange between a host computer and the on-board Z-80 microprocessor of the intelligent interface. When the host wishes to send a byte of information -- command or data -- to the interface Z-80, it loads the data lines with the appropriate values and pulses the BUSY line. This latches the data into an on-board register and sets the BUSY flag, informing the Z-80 that a new byte is in the input register. The Z-80 reads the byte, takes appropriate action, and informs the host by setting the DONE flag (automatically clearing BUSY). The host, observing that BUSY is clear, can proceed to send another byte of information. Transfer of information from the interface Z-80 to the host is roughly equivalent. The Z-80 loads its output register and sets DONE. The host then reads this register, clearing DONE and telling the Z-80 that it is ready for another byte of data. High-speed transfers in either direction can be carried out by connecting the DONE handshake to the WAIT input of the Z-80. In this way, high-speed Z-80 block I/O instructions can be used, stopping the cpu whenever DONE is set, and synchronizing the two computers. Transfer rates with such a scheme are generally limited only by the speed of the host processor.

The I/O port described above provides all communications between host and interface processors during network transactions initiated by the host processor. However, since the data network is bidirectional, it is equally likely that a transaction will be initiated by some other node of the network. In such a situation, it is necessary that the intelligent interface have the ability to get the attention of the host processor. To preserve universality, this attention-getting technique must be equally suited to both interrupt-driven and skip-check computer architectures. The service (SRVC) line of the intelligent interface fills this role. When the interface Z-80 needs to inform the host of external activity, it sets both SRVC and DONE. In an interrupt-driven computer (such as the ECLIPSE), SRVC is connected directly to the interrupt arbitration logic, and DONE is ignored. A computer using skip-check techniques (the Nicolet 1080) can monitor DONE in a skip-check loop, checking for SRVC only if DONE is set. The SRVC line can be brought in as the ninth bit of the parallel data port if host word-length permits, or it can form a single bit in a second output port. In this way the intelligent interface can perform many housekeeping chores without the knowledge or intervention of the host, notifying it only when a communications link with another device has been established.

Network: The logical link between the interface Z-80 and the data network consists primarily of an Intel 8291 GPIB Talker/Listener chip. This LSI chip is treated by the Z-80 as a peripherial device and addressed through eight 8-bit I/O ports. These ports give access to a wide variety of capabilities built into the circuitry of this chip. In addition to monitoring the BUSY line from the host computer, the Z-80 spends its free time monitoring the ports of the 8291, looking for an external network transaction request. Only after such a request is made and the necessary communications link (virtual circuit) is established, does the Z-80 notify the host computer.

Data transfer to and from the network occurs block-wise through the on-board RAM (Random Access Memory) of the interface. A block of 256 data bytes is transferred to the interface as described above and stored in the data buffer. This block of data is then transferred to the 8291 a byte at a time, until the entire block is sent. Z-80 block I/O instructions can be utilized in this transfer as well as the host transfer. The DMA WAIT line of the 8291 is connected to the Z-80 WAIT line, causing the Z-80 to execute its block I/O instructions at full speed whenever the 8291 is ready to accept a byte of data. Since the 8291 and the network itself is faster than the Z-80 in this case, normal block data transfers occur at the full cpu speed of about 190K bytes/second. If for any reason the receiving node(s) could not accept data at full speed, the block transfers would slow down to match the speed of the slowest active node. The block-wise orientation of data transfers allows all checksum generation and verification to be handled by the interface itself. The host computer need never be concerned about the intergrity of the data it sends or receives, because these functions are all handled transparently by the intelligent interface.

WARPATH Network Protocols. Data transfers on the WARPATH network are effected by a limited number of single-byte "control tokens" that are passed between host computers at two network nodes, intelligent interfaces at different nodes, or between a network interface and its host at a given node. Data transmission itself is optimized for long messages, as is often the case in nmr experiments. Each network message is broken into a number of 256 byte blocks, as mentioned above. This structure is compatible with the format of Z-80 block move instructions. A minimum message is one block, and up to 256 such blocks can be sent in a single message for a maximum length of 65,536 total bytes. Longer messages must be sent in more than one network transaction. The only restricton on message content imposed by the network interface itself is that the first byte of the first block of each message must contain the number of blocks to be sent in that message.

For purposes of protocol simplicity, a given network node can initiate a data transfer only from itself to another node. It cannot initiate reception of a data file. Thus, if a node wants to **receive** a data file, it must first **send** a request for that file to the node where that file resides. This second node then initiates another network transaction to send the file to the requesting device.

Illustration of the operation of WARPATH protocols can best be done by following a hypothetical transfer through the network. No attempt is made to demonstrate all of the error traps built into the WARPATH in this example, since this would only serve to confuse the overall pattern of data flow. The general data path is illustrated in Figure 5.10.

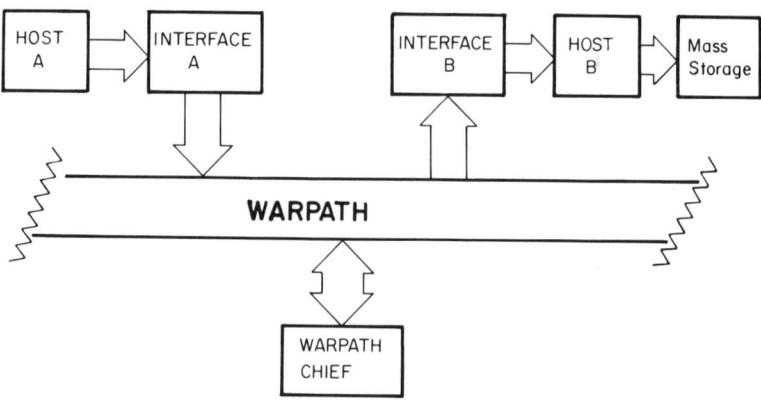

FIGURE 5.10.   *Data flow in a typical WARPATH communication. Host A sends the data to intelligent interface A which takes care of all low-level protocols in communicating with the CHIEF and node B. Intelligent interface B receives a block of data and passes it on to host B where it is eventually stored on B's mass storage device.*

The operator at network node "A" wishes to store a data file on the mass storage device associated with network node "B". This data consists of 4096 three byte data words, for a total of 12K data bytes. Associated with this data is a 1K byte file header, containing parameters associated with the collection of this data, a file name, user comments, and other relevant information. Total file length is 13K bytes, or 52 blocks. The operator initiates a network transfer by issuing a network "send" command to host A, specifying node B as the receiving device. Host A then posts a one-byte "send" control token to its intelligent interface. The interface issues a network service request which includes the address of node B. Meanwhile, the network CHIEF polls all nodes of the network in search of a device that is requesting service. When the CHIEF observes a service request from node A, it notifies B of a pending data transfer. The interface at B responds by entering its receive subroutine and notifying its host via the SRVC line that a transfer is forthcoming. After establishing a receive link at node B, the CHIEF informs node A that data transfer can begin. If node B had been inactive, the CHIEF would have informed node A and aborted the transfer. Once the CHIEF enables a data transfer, it "eavesdrops" on the message to insure that it has begun properly. After satisfying itself that everything is in order, it ignores the message, waiting only for an end-of-message signal.

Interface A, having received the go-ahead from the CHIEF, uses SRVC to request the first data block from host A. It transfers this block to the WARPATH, appending an 8-bit checksum at the end. On completion of each data block transfer, the interface again interrupts the host, requesting the next block of data, until all blocks have been sent.

While interface A is sending data blocks, interface B is receiving them and calculating its own checksums. For every data block correctly received, interface B interrupts its host and passes on the block. If a block is received incorrectly, interface B remembers the block number, receives the remainder of the file, and discards it without passing any data on to host B.

The last byte of the last block of data includes an end-of-message signal. This signal is detected by the CHIEF as a signal to reverse the direction of data exchange on the WARPATH. Node A then becomes the listener and B becomes the talker. If the entire file was received correctly, interface B waits for a control token from host B which contains information to host A regarding the status of the transferred file. Examples include successful completion, disk write errors, disk full errors, or a variety of other messages that can then be relayed to the operator at node A. If a transmission error was detected in any block by interface B, it issues a RETRY control token at the end of file transmission which is then utilized by the CHIEF to once again reverse the direction of data flow. This token is also passed from interface A to host A, requesting that the file be retransmitted in its entirety. Data transmission proceeds as before with interface B discarding all data that was previously read correctly. The receiving node can issue as many RETRY tokens as are necessary to complete correct file transfer, but an upper limit of 5 tries is fixed in software on the assumption that more errors than this in a single file must be due to a hardware failure of some type. In many cases such failures will be detected by the CHIEF and displayed on its Video Monitor.

The error recovery technique described above requires that each file be completely retransmitted when an error is detected.

At first, this appears highly inefficient. However, the extremely low inherent error rate of the WARPATH and the simplicity of implementation of such a recovery scheme argue strongly in its favor. Further, such a scheme guarantees that the receiving host only accepts error-free data, and the transmitting host only sends complete data files. This greatly simplifies the structure of the host-resident software, an important factor since this software is unique for each type of computer in the network.

As mentioned earlier, the burst transmission rate for a single block of data on the WARPATH approaches 190K bytes/second. Including the overhead associated with checksum generation and host-interface interactions, data throughput for an entire file can be as high as 50K bytes/second. Thus, the data in the example given above can be transmitted in roughly a quarter second. On completion of a given data transfer, the WARPATH CHIEF polls all other network nodes in a round-robin fashion for additional network service requests. This guarantees that no individual network node can monopolize the network for more than a few seconds, and that each node of the network gets an equal opportunity to utilize network resources.

The WARPATH local area data network, by capitalizing on state-of-the-art developments in intelligent hardware, provides a powerful and versatile vehicle for data exchange and resource sharing in a multiple-computer laboratory environment.

THE "INDIANS" ON THE WARPATH

Overview. The FSUNMR WARPATH data network, like any good warpath, would be of little use without a few Indians. In this case, the INDIANS are INteractive Data Initialization and Acquisition Nodes. These nodes are S-100 (IEEE-696) based microcomputers designed to interact with the scientist in initializing an NMR experiment; with the spectrometer for real-time experimental control and data acquisition; and with the other nodes of the WARPATH network to exchange data, programs and network bookkeeping information. The basic components of an INDIAN data station are shown in Figure 5.11.

The choice of an industry standard interconnection scheme such as the S-100 bus eliminates a host of basic and time-consuming decisions on mechanical and electrical specifications and allows more immediate concentration on overall performance characteristics of the hardware/software system. In addition, a standard bus gives access to a wide variety of competitively priced products. This is particularly true of the S-100, since no single vendor dominates the market and since the products,

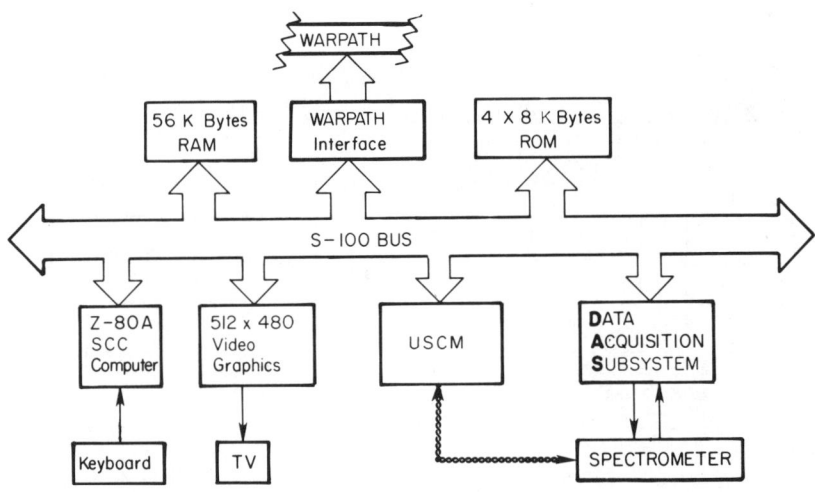

FIGURE 5.11.   Block diagram of the S-100 based INDIAN data station showing all major functional units.

geared for the computer hobbyist market, are often substantially less expensive than their industrially oriented counterparts.

The card cage and power supply currently in use for the INDIAN is the 21-slot Cromemco Z-2, but any 12-slot card cage and a power supply with at least half the available current of the Z-2 would be adequate for a complete INDIAN system.

INDIAN / Spectrometer Interface.   The most important function of the INDIAN data station is its interface with the nmr spectrometer. The demands of the FT nmr experiment in terms of timing constraints, accuracy and stability are probably as stringent as any other experimental technique found in the chemical laboratory. To meet these demands, a sophisticated S-100 based Data Acquisition Subsystem (DAS) has been developed at FSU. This subsystem is modular in design, with a minimum configuration consisting of two S-100 circuit cards. Three other boards can be added individually to increase the power and capabilities of the overall system, resulting in a maximum DAS consisting of five S-100 cards. Figure 5.12 illustrates a fully implemented DAS. The minimal system consists of one Analog/Digital Converter card and one Signal Averager card. Another A/D card can be added to permit two channel (quadrature)

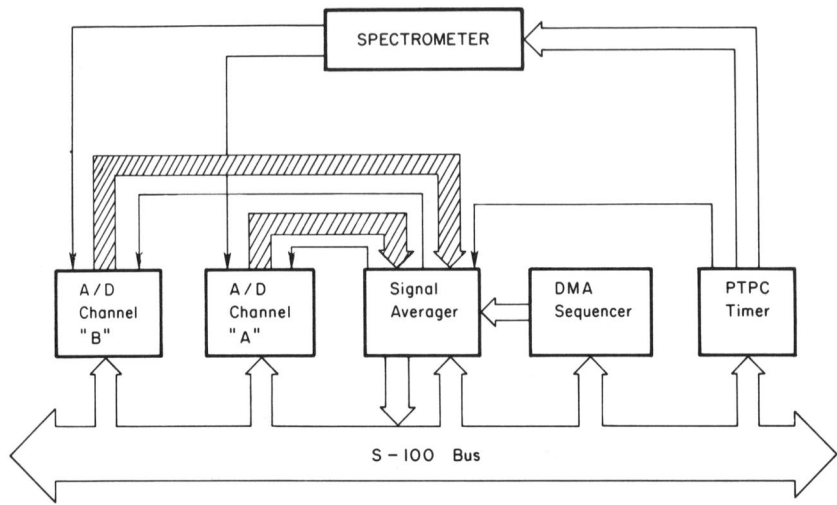

FIGURE 5.12. *The 5-board DAS. A minimal system consists of the Signal Averager and one or two A/D cards. The PTPC and DMA Sequencer cards can be added as necessary to form a more powerful Data Acquisition Subsystem.*

data collection often used in FT nmr experiments. Additionally, a Programmable Timer / Pulse Controller (PTPC) circuit card can be added, to effect highly precise timing in the data collection sequences, and a Direct Memory Access (DMA) Sequencer can be added for hardware controlled high speed signal averaging independent of the host microprocessor. This subsystem was developed specifically for FT nmr applications, but by meeting the strict demands of this environment, it can be made applicable to almost any other laboratory experiment as well.

In addition to needs met by the high performance data acquisition subsystem, it is desirable and often vital for an INDIAN data station to have some means of interacting on a lower level with the spectrometer console. Some functions such as decoupler level control and radiofrequency pulse phase shifting must be done under computer control. Other functions, including radiofrequency settings, sample temperature monitoring, and magnet shim control may be carried out by direct operator interaction, but are certainly amenable to computer control. Another S-100 subsystem under development at FSU, the Universal Serial Controller / Monitor (USC/M), consists of a single S-100 card capable of supporting as many as 255 external devices; this can fulfill any or all of the functions mentioned above. These six circuit cards comprise the hardware interface between the INDIAN data station and a spectrometer. Each of these circuits is discussed in detail below:

   Analog-to-Digital Converter: The linch-pin of any computerized analog data acquisition scheme is the analog-to-digital converter. A wide variety of A/D components are available from S-100 vendors, but most are designed for versatility of input types, or with a multiplexed conversion unit for a large number of non-simultaneous analog inputs. The FT nmr experiment normally has only one (singlature) or two (quadrature) analog signals of prime importance: the heterodyned audio frequency response from the receiver circuitry of the spectrometer. This response is very critical, however, and often requires extensive signal conditioning before being digitized. Further, the most common approach to the quadrature experiment requires that both analog channels must be sampled simultaneously. Multiplexed A/D circuitry is generally not fast enough in such a situation, and would necessitate 2 sample-and-hold circuits to preserve exact simultaneity. In the case of the DAS, this has led to the development of an A/D conversion scheme in which a complete and independent analog channel is implemented on a single S-100 circuit card. Two such cards can then be employed for the required simultaneous data channels of a quadrature detection scheme.
   The circuitry on the A/D Converter board can be divided into two general functions: the signal conditioning function, and the sample and conversion function. A block diagram of both aspects of the A/D converter circuitry is shown in Figure 5.13.
   In the signal conditioning portion of this circuitry, an analog voltage is brought into the circuit via a co-axial cable. DC voltage offset is removed from this signal automatically by a negative feedback integrating operational amplifier. This automatic loop can be defeated by a software switch, allowing the operator to remove DC bias manually by controlling the setting on a potentiometer. The analog voltage is then amplified by a computer-controlled variable gain amplifier. The six available gain settings range from X1.0 (unity gain) to X12. After proper amplification, the signal passes through a 4-pole Butterworth low-pass filter to remove high-frequency noise components. This filter is a digitally controlled hybrid module with a maximum bandpass of 20 KHz. It can be replaced during assembly with a 50 KHz module having identical pinouts, if so desired. The programmable cutoff frequencies for the 20 KHz module range from 80 Hz to 20 KHz in steps of 80 Hz over its entire range.
   The sample and conversion functions of this card are carried out by another hybrid module containing both sample-and-hold circuitry and A/D conversion circuitry (the Datel ADC-HS12B). On

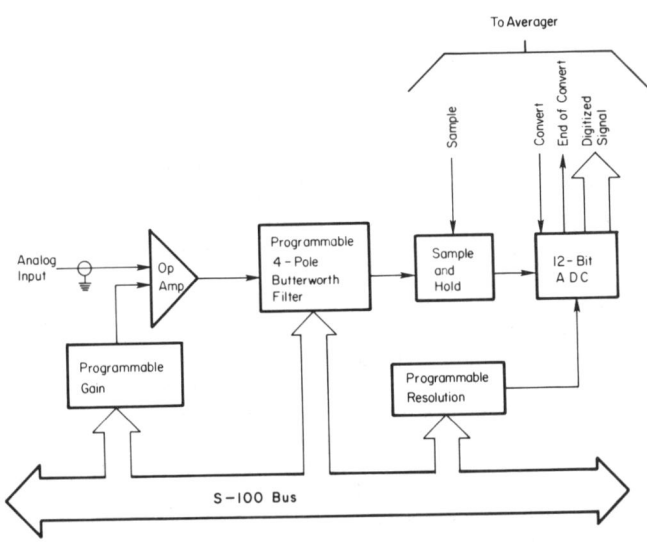

FIGURE 5.13. Major features of the DAS A/D Converter include gain, low-pass filter, and digital resolution, all under software control. (Control signals are provided externally from the Signal Averager circuit.)

command, this module shifts from sample to hold modes, and causes digitization to begin. An end-of-convert signal is generated on completion of the conversion. The A/D module has a maximum resolution of 12 bits (1 part in 4096) ±0.5 bit. A 12-bit conversion takes 8 microseconds resulting in a sampling limit of 125 KHz, or a frequency window of 62.5 KHz for single phase detection (±62.5 KHz for quadrature detection). The resolution of the A/D converter is software selectable from 12 to 6 bits in two bit steps. Higher data rates can be achieved by selecting lower resolution. For example, 6-bit conversions require half the time of 12-bit conversions, resulting in a frequency window of 125 KHz (substantially faster than can be handled by the computer!) The digital outputs from the A/D converter are not brought to the S-100 bus. Instead, they are buffered through tri-state drivers onto a 26-wire ribbon cable, making them available to the third of the 5-board subsystem, the Signal Averager.

Signal Averager: NMR is inherently insensitive when compared to various other spectroscopic techniques. To compensate for this lack of sensitivity, signal averaging (actually just signal summation) is often employed[1,2]. Since a maximum 12-bit resolution is available from the A/D hardware, it is obvious that at least two words of an 8-bit microcomputer are required to effect any signal averaging. However, only sixteen coherent 12-bit signals could be co-added into a 16-bit word (16 bits - 12 bits = 4 bits; 2^4 = 16 scans) before overflowing the range of that word. Thus a logical data word length for such signal averaging becomes 3 bytes, or 24 bits. The FSU nmr Data Acquisition Subsystem was designed to utilize both two byte (16-bit) data words for speed and efficiency, and three byte (24-bit) data words for high resolution. The speed limitations of 8-bit microcomputers becomes painfully evident when one attempts to write a software loop for efficient 24-bit signal averaging. Such a loop, in assembly language, can easily take as much as 150 microseconds per channel to execute. This puts an upper limit on sampling bandwidths of 3.3 kHz, much more restrictive than the A/D hardware and too restrictive for many FT nmr experiments. The bandwidth of the experiment is dramatically improved if the signal averaging function is relegated to hardware instead of software. Hardware signal averaging can be done in two stages with the DAS. The first stage is discussed here; the second is covered in the section on the DMA Sequencer.

The Signal Averager board, functionalized in Figure 5.14, is the hub of the DAS. It provides the link between the digitized output of the A/D boards and the microcomputer. It also coordinates the activities of the remaining two circuit boards in the DAS: the Programmable Timer/Pulse Controller, and the DMA Sequencer (if either of these boards is used). The Signal Averager can be divided into three functional areas: the signal averager itself, the programmable interval timers, and the I/O control logic.

The signal averaging portion of this circuit consists of a 24-bit adder and a 24-bit data register. The low-order 12 bits of the adder can be electronically connected to the 12 data input lines from either of two A/D boards under software control. The adder can then either add or subtract this data into the contents of the 24-bit register. The data register must be loaded with the data word to be averaged (either 24- or 16- bit) prior to this operation, and unloaded when averaging is completed. If the loading and unloading is done by the Z-80, using its block I/O instructions, it can be done in about 95 microseconds per pair of 24-bit quadrature points. When higher speeds are required, the same procedure is performed by the DMA Sequencer card in roughly 12 microseconds.

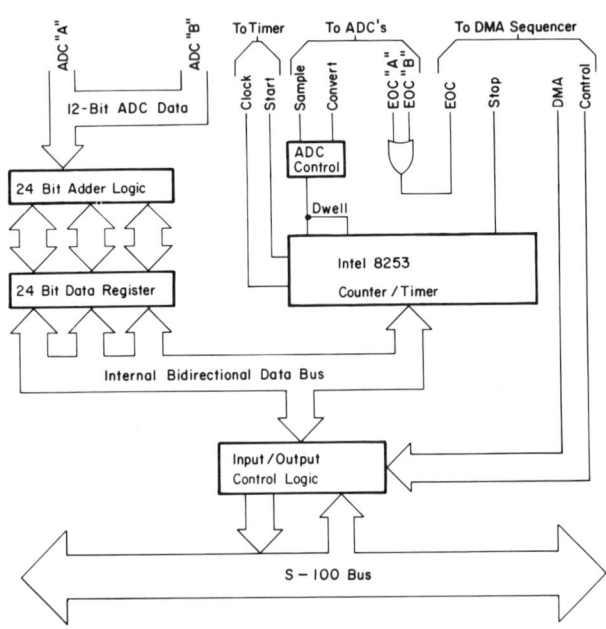

FIGURE 5.14.   *The DAS Signal Averager generates timing and control signals for one or two A/D Converter cards, and adds digitized data to computer memory under either CPU or DMA control. The timing is derived from either the 2 MHz S-100 system clock, or from the 20 MHz PTPC Timer clock.*

Three counter/timers are employed on the Signal Averager board in the form of a single Intel 8253 Programmable Interval Timer integrated circuit. Each of these three counter/timers is a 16-bit programmable counter with a maximum clock rate of 2 MHz. The first of these counters is used as a dwell clock, generating the sample and convert signals that control the A/D boards. Its 2 MHz clock rate is derived from the S-100 system clock or, if present, from the 20 MHz clock of the PTPC circuit board. This dwell clock has a resolution of 500 nanoseconds and is programmable to generate dwell times ranging from 1 microsecond to slightly more than 32 milliseconds. A second counter is employed to count the number of data points in the data array. The Fast Fourier Transform algorithm utilized in FT nmr usually requires that the number of data points be a power of two, but the array specified by this counter can be any size from 1 point to 65,536 points. By keeping track of the data array count in hardware, the software overhead is greatly reduced, consequently increasing data throughput. The third counter of the 8253 circuit is used as a scan counter, keeping track of the number of times data has been signal averaged into the array. This feature once again reduces the complexity of the necessary software.

The I/O control logic is the decision center of the Signal Averager. This logic decodes a number of input and output commands that provide the interface to the S-100 bus. One such input command allows the computer to monitor the DAS, determining which circuit boards are resident in a given INDIAN system, and allowing the software to adapt itself to the hardware environment without any external reprogramming. Additionally, this circuitry enables the computer to determine whether or not the dwell clock is running, and enable or disable automatic start, automatic stop, and other functions.

Three I/O commands allow the loading and unloading of the 24-bit on-board data register. When the DMA Sequencer is active, these commands are disabled and the data register contents are controlled by signals from the DMA circuitry.

<u>Programmable Timer / Pulse Controller</u>: Highly accurate and reproducible timing sequences are often required to prepare the nuclear spins for data acquisition in the FT nmr experiment. Also, control of these sequences must be highly flexible to accommodate the multitude of variations currently in use. The length of the time periods in these sequences can vary from sub-microseconds to tens or even hundreds of seconds, and they must be reproducible for periods as long as tens of hours.

Furthermore, software timing loops are not acceptable in such an application for a number of reasons. Execution times for today's microcomputers are generally on the order of a few microseconds per instruction, leading to minimum timing loops taking tens of microseconds to execute, several orders of magnitude longer than desirable. Software timing sequences must be constructed very carefully to minimize timing jitter, requiring that they be written by relatively experienced programmers. This inevitably restricts their modification, and hence their adaptability to new or unusual timing sequences. Some inaccuracies are unavoidable with software timing, due to factors such as conditional branching. These inaccuracies become intolerable when, as is often necessary, techniques such as cycle-stealing DMA and interrupt-driven peripheral handling are employed to optimize the microcomputer environment.

These and other considerations led to the development of the Programmable Timer/Pulse Controller (PTPC) for the INDIAN Data

Acquisition Subsystem. The PTPC is a high resolution, highly accurate and intelligent timer module, patterned loosely after other FT nmr timers[3] and implemented on a single S-100 circuit card. It has the capability of maintaining up to 16 different program steps, supports internal branching and looping, and maintains an 8-bit programmable hardware pass counter to inform the computer when a loop has been executed a given number of times. The PTPC also has sixteen buffered "pulse" and eight buffered "level" outputs that interface it with the outside world. A block diagram of the PTPC circuitry is given in Figure 5.15.

The timer memory of the PTPC is configured as sixteen timer words, each word containing 40 timer bits and 8 control bits. These words consist of independent memory and run-time latches.

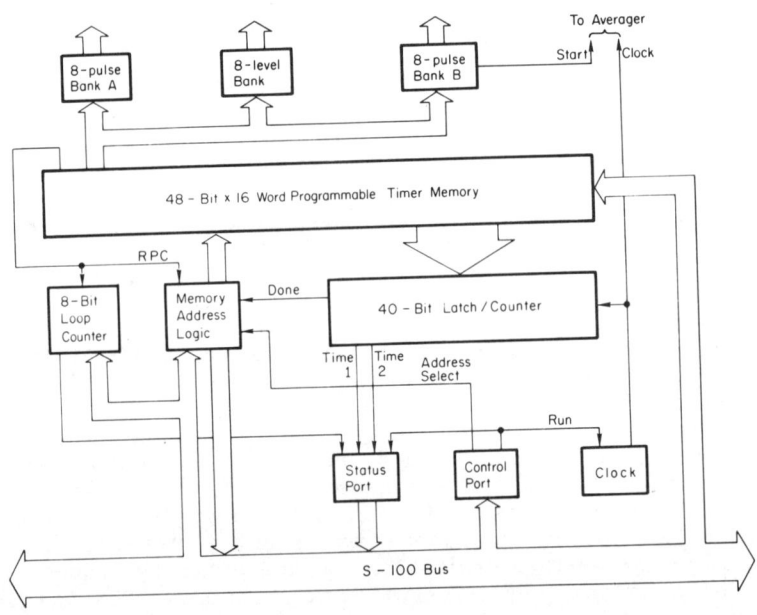

FIGURE 5.15.   The main element of the DAS PTPC is its 16-word timer memory and 40-bit latch, permitting complex and highly accurate timing sequences. In addition, up to 24 signals are available for control of external events. The 20 MHz master crystal allows time resolution of 50 nanoseconds.

This feature is important in that it allows the memory to be modified _during_ the execution of a timer sequence.

Any byte of timer memory, or even the entire timer array can be modified "on-the-fly" with no disturbance in the current program step. In this way, very long and highly complex timer sequences can be easily accomodated. The only restriction to reprogramming the timer memory is that the next timer transition cannot occur until the reprogramming is completed. However, since the entire timer memory can be reprogrammed in less than a millisecond, this poses no severe handicap in most cases. In any event, flags are available to the computer (see below) to indicate when enough time is left for reprogramming.

The 40 timer bits in each timer word and the 20 MHz crystal-controlled clock on which the timer is based, allow the direct selection of pulse time periods from 50 nanoseconds to over 15 hours in length, with 50 nanosecond resolution over the entire range. This comfortably provides the resolution, accuracy, and dynamic range required by most FT nmr experiments.

Six of the eight control bits associated with each timer word are involve with pulse and level control. In the context of the PTPC, a pulse is defined as being "on" (TTL logic 1) for the duration of the time period in which it is selected. A level, on the other hand, is turned "on" at the beginning of a given time period and remains on until it is turned "off" at the beginning of another time period. The "on" and "off" levels are specified by one of the remaining control bits of the timer word. Any of sixteen pulses and eight levels can be selected for any of the sixteen timer words, and with some limitations, up to two pulses and one level can be selected for any given timer word. These 24 output signals are brought to the computer chassis via a 50-wire ribbon cable (alternate wires provide ground returns for each signal) where they are buffered through line drivers before being routed to the spectrometer as required.

The last of the timer control bits is the Reset Program Counter (RPC) bit. If this bit is set in a timer word, it causes the internal timer address to be reset to the currently defined "first address" of the sequence. The timer program then branches to this address for its next program step, rather than executing the next sequential address. The RPC bit is also used elsewhere to count passes through a loop in a timer program.

The PTPC is directly interfaced to the Signal Averager through a 14-wire ribbon cable. This cable carries the 20 MHz clock signal, allowing an entire DAS to be synchronized to the same crystal clock. The cable also carries two other signals: a hardware start pulse, that can be used to start the dwell clock on the Signal Averager; and a signal to inform the Averager that, yes, there is indeed a PTPC in the system.

The interface to the S-100 bus is implemented via eight 8-bit output ports and one 8-bit input port. Six of the output ports are used to load the 48 bits (six 8-bit bytes) of each timer memory word. These six ports are allocated sequentially in the computer's I/O space to utilize block move instructions to rapidly load timer memory directly from computer memory. The seventh is used to load an 8-bit loop count indicating the number of times an inner loop is to be executed before informing the host computer. This count is decremented each time the counter receives the RPC signal, and when it reaches zero, the computer is informed by a bit in the PTPC output port. It is then up to the computer to either stop the PTPC, or reprogram timer memory to execute another portion of the timing sequence. The last output port to the PTPC contains several control signals: a start and a stop signal to the timer circuitry and a timer memory address switch. Four remaining bits of this port serve to select one of the sixteen timer memory addresses. These bits are saved in an on-board register and act as both the starting address of the timer program and the return address of a program loop. In addition, if the memory address switch mentioned above is "on", this memory address is directly selected, allowing it to be reprogrammed via the first six output ports. The contents of this address register, like the rest of timer memory, can be modified while the timer program is running.

The single input port from the PTPC provides monitoring information to the computer. Four of its eight bits reflect the current value of the timer program counter, allowing the computer to keep track of which step in the timer program is the next to be executed. Two of the remaining bits provide a rough estimate of the amount of time remaining in the current program, giving the computer the information it needs to decide whether or not it can reprogram all or part of timer memory. One of the last two bits in this port serves as a RUN/STOP indicator for the state of the timer clock, and the other bit provides information on the loop counter, as mentioned above.

Direct Memory Access Sequencer: The 95 microsecond cycle time offered by the Signal Averager under software control is fully adequate for the majority of signal averaging applications. This is particularly true for nmr spectrometers operating at low magnetic fields, or for nuclei such as protons, with small chemical shift dispersions. However, modern superconducting magnets can operate at very high fields. Also, multi-nuclei spectrometers such as the 3.52 Tesla SEMINOLE spectrometer constructed at FSU can often be used to observe nuclei with very large chemical shift dispersions (e.g. lead, cobalt, platinum, etc.). These applications require larger spectral windows, and hence, faster sampling rates than those achievable with the Signal Averager alone.

Even with the hardware averaging capabilities of the Signal Averager, the computer itself remains the bottleneck in data sampling rates. To significantly improve data sampling throughput, the microcomputer must be stripped of even the simple data pushing chores required of it in conjunction with signal averaging. The remaining circuit card of the DAS was designed to relieve the computer of exactly those chores. It is called the Direct Memory Access (DMA) Sequencer because it accesses system memory directly with no computer intervention, and it generates the appropriate sequence of steps necessary to complete multi-byte data transfers between the Signal Averager and memory.

The DMA Sequencer, currently in final design and testing stages, is illustrated at the functional level in Figure 5.16. It is simultaneously both the simplest and the most complex of the five DAS circuit boards. It is the simplest because it has only one basic function: moving data between memory and the

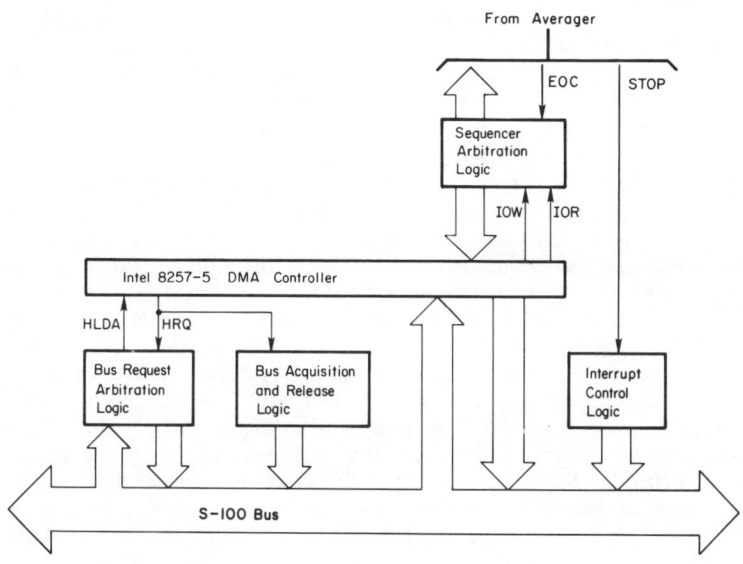

FIGURE 5.16.   *The DMA Sequencer provides an S-100 standard DMA interface and all sequencing logic to permit high speed signal averaging on one or two channels of 16- or 24-bit data words.*

Signal Averager. It is at the same time the most complex circuit in the DAS because in order to carry out this single function, it must cause the host computer to relinquish control of the S-100 bus in a non-destructive manner, take over control of all the appropriate bus signals, and finally return control peacefully to the host processor, a difficult course of events in the S-100 environment.

The DMA Sequencer can be logically divided into two functional blocks: the circuitry that interfaces with the S-100 bus, and the circuitry that connects to the Signal Averager.

The heart of the DMA Sequencer is a Large Scale Integrated (LSI) circuit from Intel, the 8257-5 DMA Controller. This chip is essentially a dumb but fast microcomputer programmed in hardware to move data back and forth between memory and peripheral devices. It can act as both a "slave" and a "temporary master" processor with respect to the S-100 bus. In the slave (or program) mode, the 8257-5 can send or receive data bytes through I/O ports on command from the host computer, much the same as for the other boards in the DAS. This is the manner in which the circuit is programmed to carry out its DMA function. During DMA operation, the chip (aided by some support circuitry) acts as a temporary S-100 bus master, as specified in the S-100 bus standard. It issues a hold request, and on receipt of a hold acknowledge from the host computer, proceeds to put the host "to sleep" in an orderly manner. On completion of its programmed DMA function, the process is reversed and the host computer continues execution of the program it was running when it received the hold request, totally unaware that a DMA cycle has occurred. As mentioned earlier, software timing loops can be adversely affected if they are done concurrent with DMA.

All of the communication between the DMA Sequencer and Signal Averager occurs through a 14-line ribbon cable connecting the two boards. One of these lines informs the Signal Averager that a DMA Sequencer is present in the DAS, allowing the appropriate software configuration to be set up. Another is active only during a DMA memory access, serving to lock out the I/O commands from the computer that control the loading and unloading of the 24-bit data averaging register. These commands are replaced with three read and three write signals generated on the DMA Sequencer board and connected through the ribbon cable. One signal on this ribbon cable travels from the Signal Averager to the DMA Sequencer. This signal, End-Of-Convert (EOC), is a pulse generated whenever both A/D Converters completes a data conversion. EOC is employed by the DMA Sequencer to initialize operation of the sequencing circuitry (if DMA is enabled). Two remaining signals, DACK0 and DACK2, are used to enable the outputs of one or the other of the two A/D Converter cards in the DAS.

Four possible data byte routing sequences are software selectable in the DMA Sequencer hardware. These four sequences correspond to: single phase 16-bit, single phase 24-bit, quadrature 16-bit and quadrature 24-bit data averaging schemes. Each of these schemes requires that a different sequence of I/O signals be sent to the Signal Averager board. For purposes of illustration, the quadrature 24-bit sequence is discussed below.

The 8257-5 DMA Controller chip has four independent DMA channels associated with it. Each channel can be programmed to either read from or write into computer memory, and each channel maintains its own memory reference address. In quadrature DMA operation, the sequencer employs all four of these channels. Channel 0 is programmed to generate the signals which read data from the memory array containing signal B, channel 1 is utilized to write the averaged signal back into this array, while channels 2 and 3 perform equivalent functions for memory array A. With these four channels properly programmed, the computer need have no further interaction with the DMA circuitry during signal averaging until an entire scan through the data array is completed.

To review the way in which the separate circuits of the DAS work together to acquire and average data, and to illustrate the function of the DMA Sequencer, the following discussion will lead the reader stepwise through the complete data collection sequence for quadrature 24-bit nmr data:

Initialization of data collection requires the programming of each of the cards in the DAS. The A/D circuits must be programmed identically with respect to filter settings, analog gain, and digitizer resolution. The Signal Averager must be programmed with the proper dwell time setting, the correct data array count, and the scan count. The computer must load the PTPC with the timing sequence that controls this specific experiment. Finally the DMA Sequencer must be configured for quadrature 24-bit operation and loaded with the starting addresses for both the signal A and signal B data arrays.

After all of the subcircuits of the DAS have been properly programmed, data acquisition is begun by issuing a START instruction to the PTPC board. Following this, the host microcomputer is free to execute whatever housekeeping or I/O management programs are necessary to properly maintain the computer/operator interface.

The PTPC START causes the timer to begin "ticking". At some later time, this generates a hardware start signal from the PTPC to the Signal Averager, initializing a scan through the data arrays. The dwell clock on the averager board is turned on by this hardware start, periodically generating conversion signals to the A/D Converter cards. Following each convert signal, both the A/D cards respond at roughly the same time (since they started at the same time). This response is converted into an EOC pulse and transmitted to the DMA Sequencer.

The EOC signal triggers the sequencing circuitry and causes it to generate a DMA request on channel 0 (DRQ0). As this is the first in a sequence of DMA requests, the DMA circuitry must establish control of the S-100 bus by issuing a hold request signal to the host computer. The computer responds within a microsecond, regardless of what program it is executing, by issuing a hold acknowledge signal. At this point, the DMA circuitry has control of the bus and will maintain control until this entire DMA sequence is complete.

The DMA circuitry responds to a hold acknowledge with a DMA request acknowledge (DACK0) which is used to enable the outputs from A/D circuit B. Since channel 0 has been programmed as a "read" channel (see above), an I/O read strobe is generated by the 8257-5 and decoded by the sequencing circuitry as the command that reads a data byte from memory and loads it into the low-order portion of the 24-bit data register on the Signal Averager board. This completes the first cycle of the sequence. The sequencer continues to assert DRQ0 at this point, and since the DMA circuit is still in control of the bus, it can rapidly respond by issuing another I/O read strobe. This second strobe is decoded by the sequencer to cause the next data byte from memory to be loaded into the middle byte of the 24-bit data register. This step is repeated once more, loading the high-order byte of the averager register. On completion of the third byte transfer, the sequencer shifts the DMA request signal from DRQ0 to DRQ1.

DMA channel 1 was previously defined as a "write" channel, so the next three steps of this sequence serve to move the averaged data from the low, middle and high bytes of the data register back into their original memory locations. The sequencer again shifts the DMA request, moving it from DRQ1 to DRQ2, generating DACK2 and enabling the data outputs from A/D circuit A. These outputs are then averaged with the contents of memory array A in a manner identical to that described above for memory array B.

When the last write strobe is generated and the sequencer shifts the DMA request out of DRQ3, no further DMA requests are pending, and the DMA controller relinquishes bus control to the host computer. This entire sequence, although long in the telling, requires only 12 microseconds to execute. Two 24-bit words from two separate locations in memory have been added with two 12-bit outputs from two independent A/D converters, and replaced in memory. All of the addresses in the DMA controller have been advanced by three locations and are pointing to the next word in each array. The sequencer circuitry has returned to its initial state, ready for the next EOC signal. The computer is unaware that it has lost a few machine cycles, and continues executing its housekeeping programs.

The dwell clock has continued to tick during all of this frenetic activity, and after each dwell period, it generates another convert signal, causing the entire sequence to be repeated. This cycle continues until the array counter in the Signal Averager circuitry reaches its preprogrammed value. The dwell clock is then automatically halted and the host computer is interrupted with an end-of-scan interrupt. The host must reset the addresses in the DMA channels to the starting points of their respective arrays, and determine whether or not the requested number of scans has been completed. If not, it returns to its housekeeping chores, waiting for the next end-of-scan interrupt. If so, the computer issues a STOP command to the PTPC, disables the DMA circuitry, and exits the data acquisition subroutine. The host computer must be able to complete this reprogramming and decision-making between the time the last data point in an array is averaged and the time when the PTPC issues the next hardware START instruction to the dwell clock on the averager board, requiring a built in time delay on the order of one hundred microseconds between the end of one scan and the beginning of the next.

The DMA Sequencer and the other circuit boards of the FSU DAS represent a major investment in hardware design and implementation. A modular approach was utilized in their development, resulting in two distinct advantages: functional subsystems can be implemented with only the circuit elements required for that specific environment, minimizing both system costs and complexity; also any individual card in the system can be modified or redesigned to change operating characteristics or to take advantage of newer technologies, as long as the interfaces between cards remains unchanged. These advantages coupled with the inherent power of the DAS subsystem make feasible the implementation of microcomputer-based high performance nmr.

<u>Universal Serial Controller</u> / <u>Monitor</u>: FT nmr, like many experimental techniques, has special requirements that must be met by specialized hardware such as the FSU NMR Data Acquisition Subsystem. But, also like many other techniques, a wide range of functions require only simple low-speed monitoring and interaction. These functions are often handled external to the computer, because direct operator control is usually simpler to implement, and because direct computer control has often implied a maze of wires connecting computer and instrument, making both devices highly specialized and non-standard. Complex wiring makes it very difficult to update or expand the computer/instrument interface without major redesign and/or

recabling efforts. Additionally, it seems in these approaches that there is invariably one more function to computerize than there are wires available for interfacing.

The aforementioned difficulties are circumvented if an approach is taken which places some intelligence in the instrument itself, independent of the controlling microcomputer. Such an approach becomes feasible through the use of modern LSI single-chip microcomputers. One device, the Serial Control Unit (SCU-1) is currently available as the Mostek 14007 for under thirty dollars. This device is a pre-programmed microcomputer designed to perform exactly the kinds of functons described above. One such device can, with appropriate support circuitry, conceivably provide for all the low-level computer interfacing requirements of a given instrument. However, the real power of these devices lies in the fact that up to 255 of them can be connected to a controlling computer via a _single_ serial line. This feature, coupled with the low cost of the device makes it possible to incorporate several SCU-1 chips in a complex instrument avoiding a great deal of support circuitry and complicated wiring.

The SCU-1 forms one half of a Universal Serial Controller / Monitor (USCM) network currently being implemeted for the SEMINOLE spectrometer at FSU. Initially, SCU-1's are being used for essential functions such as decoupler programming and rf phase control, and plans are underway to utilize them for frequency synthesizer programming, sample temperature control, and other non-critical functions.

Interfacing a microcomputer to a network of SCU-1 devices is straight-forward. The SCU-1 is a passive device, responding only when addressed by the host computer. It accepts a standard 8-bit serial data format such as that used for computer terminals and could thus be easily driven by commercially available S-100 UART (Universal Asynchronous Receiver Transmitter) circuitry over standard RS-232 compatible lines. However, since the SCU-1 uses +5 Volts, and the RS-232 standard requires ± 12 Volts, this would entail two extra voltage sources at each SCU-1 device.

A two stage approach is being taken at FSU to develop the computer-based portion of the USCM network. The first stage involves the development of a very simple S-100 circuit card containing I/O port decode circuitry, a UART chip, and a differential RS-422 interface. This interface operates on +5 Volts only, and the differential drivers offer better noise immunity than RS-232, making it an excellent choice for use in the USCM. In this stage of development, the host computer interacts directly with the SCU-1 devices, polling each one to monitor status or to output control signals. As more SCU-1 devices become incorporated into the spectrometer and hence more polling becomes necessary to monitor system status, the overhead on the host computer could become substantial.

The second stage implementation of the USCM will alleviate this overhead by distributing more intelligence onto the USCM/S-100 interface card. At the second stage, the USCM card will contain its own microcomputer, program memory, and data memory. It can then accept commands from the host computer at much higher speeds than the network can accept them, eliminating much of the time that the host computer would spend in USCM communications. Additionally, this card can be programmed to monitor any number of SCU-1 devices, checking that signals lie within predefined boundary conditions, and notifying the host computer only if these boundaries are exceeded.

The USCM is yet another example of the ways in which the proliferation of microprocessors and distributed processing techniques can radically alter the approaches and solutions to a wide variety of problems.

INDIAN / Operator Interface. In addition to providing a flexible spectrometer interface, an INDIAN data station must also provide a facile human interface. The most powerful system available will remain under utilized unless careful attention is paid to the ways in which the operator interacts with his or her computerized environment. The most obvious aspect of this environment is the program or software that allows the operator to manipulate the spectrometer. But even more basic than the software is the hardware or circuitry supporting it and providing a vehicle through which the software and the operator can communicate.

The hardware of the Data Acquisition Subsystem is discussed in some detail in the previous sections; much of the remainder of the INDIAN hardware is discussed below, followed by a description of the INDIAN software environment.

Since the circuitry that comprises this portion of the INDIAN is all commercially available, it is not discussed in great detail. Salient features are pointed out if they are important and/or peculiar to this application.

Computer and Memory: At the center of any microcomputer system is the computer itself. In the case of the INDIAN, this is a 4 MHz Z-80A, implemented on a Cromemco Single Card Computer (SCC). The SCC is a complete computer on-a-card providing: the Z-80A itself, a serial (UART) interface, several parallel interfaces, 1K byte of Random Access Memory (RAM) and 8K bytes of Programmable Read Only Memory (PROM). Although it is a complete computer, little contemplation is necessary to realize that the memory provided by the SCC is woefully inadequate for either high resolution data acquisition or a sophisticated nmr operating system. To rectify this situation, external memory has been added in the form of both RAM and PROM circuit cards.

The RAM memory is contained on a single S-100 card which supports 64K bytes of dynamic memory in the form of 16K-bit integrated circuit chips. The memory refresh, a problem with some dynamic memories, is handled on-board -- transparent to both CPU and DMA access cycles. The RAM memory can be disabled and effectivly removed from the memory map in 4K-byte blocks by hardware switches on the board. This is important in that 64K bytes will occupy the entire address space of an 8-bit microcomputer like the Z-80, and "holes" must be created in this memory address space to allow room for PROM memory and other memory-mapped devices.

As described earlier, the DAS hardware is configured to support one or two channels of high-resolution data, with word lengths in each channel of either 16 or 24 bits (2 or 3 bytes). Thus, the largest data array that can fit in 64K bytes of memory consists of 16K words in one channel, or two 8K word arrays, in each case using a total of 16K*3 or 48K bytes. This leaves 16K bytes of memory remaining for the nmr operating system and all associated parameters and variables. Although an adequate operating system could be written to reside entirely in this memory space, a hard upper limit on program size at such an early stage of development places severe restrictions on future program flexiblity and expansion capabilities. In light of this and the hardware limit of 64K bytes of memory space, some form of virtual memory allocation is indicated.

One possible virtual memory approach is the maintainance of program overlays on a mass storage device such as a disk, writing them into memory only as they are called. Such an approach has been taken, for example, in the FSUNMR software described later in this chapter. This type of approach is unsatisfactory in the case of the INDIAN, since its prime mass storage device is the WARPATH data network which is inherently slower than a directly attached disk. Also, it is undesirable to force either data acquisition or operator interaction to be directly reliant on specific nodes of the data network for program execution.

The problem of limited program space can also be solved by memory mapping techniques in which the entire program is resident in memory, but only a fragment of that memory is "mapped" into the active memory address space at any given time. The major advantage of this method is the independence afforded to each INDIAN data station. Program execution is not tied to any external device, and, like memory mapping itself, overlay execution occurs much more rapidly than mass-storage interactions. This is the method of choice for the INDIAN data stations.

The "mappable" portion of memory for the FSU INDIAN resides in one-half of the 16K bytes allocated for program memory. This memory is implemented on a Cromemco 32K Bytesaver PROM memory card, modified to support four independent 8K banks of memory on a single S-100 circuit board. Up to six such boards can be accommodated in the system, allowing a total program length approaching 200K bytes, sufficient for _any_ envisioned future expansions. The standard INDIAN memory configuration consists of one such PROM card in addition to the previously mentioned RAM memory, allowing up to 40K bytes of PROM (including 8K from the SCC card; see above) in addition to 8K bytes of program RAM and 48K bytes of data RAM.

The decision to implement INDIAN software in PROM memory carries with it a number of advantages and disadvantages. The primary advantage is once again embodied in the resulting independence of each data station. With this configuration, an operator can do instrumental check-out, experimental set-up, and data collection totally independent of all other network nodes, relying on the WARPATH only for data storage at the end of an experiment. It is also possible, with addition of local mass-storage or hardcopy facilities, to reconfigure the INDIAN as a stand-alone low-cost computer for any nmr spectrometer. Another advantage of PROM memory lies in its modularity. New firmware (software-in-PROM) can be added to the system in 2K-byte increments. This allows simple system expandability and makes the memory highly cost effective, since memory chips are added to the system only as required to contain all of the software.

The prime disadvantage of a PROM-based software system lies in the programming of the chips. Since the program becomes part of the hardware, modifications require the physical removal and reprogramming of each affected chip in every active system. Programs and subroutines must be thoroughly tested in RAM memory before being committed to PROM. This inconvenience can be alleviated to a large extent by modular programming techniques in which modifications to one subroutine have little or no effect on the remainder of the program.

Video Graphics Interface: The INDIAN presents its information to the operator on a standard raster scan TV monitor. A high resolution video graphics board (the MicroAngelo from SCION Corporation[4]) is utilized to provide video graphics and alphanumerics with a resolution of 512 (horizontal) by 484 (vertical) picture elements. The video circuitry is actually a microcomputer in its own right, consisting of a Z-80A microprocessor, up to 8K bytes of PROM memory (4K bytes of which contain dealer supplied video control software), and 32K bytes of

RAM memory, all implemented on one S-100 circuit card -- totally independent of the system memory and computer. Most of the RAM memory is used to store the video picture elements, but 1.5K bytes are available for user supplied subroutines, alternate character sets, or buffer areas for high speed video updates.

The intelligence of this video display circuitry makes possible a multi-processing approach to the video display software. The host computer can transmit data at high speed to the video computer and attend to other tasks such as monitoring the DAS hardware, the spectrometer, or user input from a keyboard, while the video computer is busy updating the display. Several such applications are employed in the INDIAN software.

The INDIAN Software Environment: When power is applied, the INDIAN data station executes the program residing in PROM on the SCC board. All other PROM banks in the memory map are disabled. This 8K bank is thus inherently different from any other bank in the system. It is subdivided into four relatively independent subprograms.

The Initialization program, executed first, serves to initialize all of the hardware of the INDIAN to a non-invasive, non-destructive passive state. On completion of this task, it presents the video display shown in Figure 5.17 and waits for operator input from the keyboard. At this point the operator has access to any of the other three programs residing in on-board SCC PROM memory. A special sequence of key presses, known only

FIGURE 5.17. Photo of the log-on display for the FSU INDIAN. From this point, the user can access the Conversational Command Interpreter or one of two diagnostic programs.

to (or at least utilized only by) systems level operators, allows entry to two restricted access programs. One of these, the machine code monitor, permits its user to examine and modify the binary contents of memory, cpu registers, and I/O ports. This program is useful for trouble shooting software or hardware problems and debugging new software. The other restricted access program contains various diagnostic routines designed to test the status of various portions of the INDIAN hardware. These routines can be run periodically to verify system functionality, or in the case of a hardware failure, can be useful in identifying the faulty component.

Any key press other than the special sequence mentioned above will result in entry to the Conversational Command Intrepreter (CCI). This program is the supervisor that controls the video display and interfaces the operator with specific applications commands. In the context of this disscussion, the application is nmr data collection. However, the CCI is designed as a general command processor for any data acquisition function and can be adapted to a new application by replacing the nmr specific PROM modules in the mapped memory banks with those of the new application.

On entry to the CCI program, a log-on subroutine is executed. This routine serves to notify the supervisory node of the WARPATH data network that a new user is beginning operation. The user identification and time of day is then recorded for bookkeeping and billing as required. If the supervisory node of the network is not active, or if the network itself is either not in use or not functional, the operator is so informed. Thus no software modifications are necessary for an INDIAN to be used as a stand-alone device without a network interface. Before transferring control to the Command Interpreter, the log-on routine moves the CCI machine code from the SCC PROM space, where it is part of the mapped region of memory, to program RAM where it can supervise the remapping function as necessary.

The video display is designed to be similar to that of the FSUNMR processing program running on the ECLIPSE mini-omputer. In this way a user suffers minimal visual disorientation in shifting from one program environment to another. The overall features of this display, as shown in Figure 5.18, include a 512 X 256 data display area in the top half of the screen, a user interactive (terminal emulation) area in the lower left quadrant, and a computer prompting and help area in the lower right quadrant. The display of data and alphanumeric information is handled by special subroutines added to the video computer's repertoire that serve to greatly reduce the work load of the host computer in this display environment.

The Conversational Command Interpreter is designed to be friendly without being pushy. Four levels of user interaction

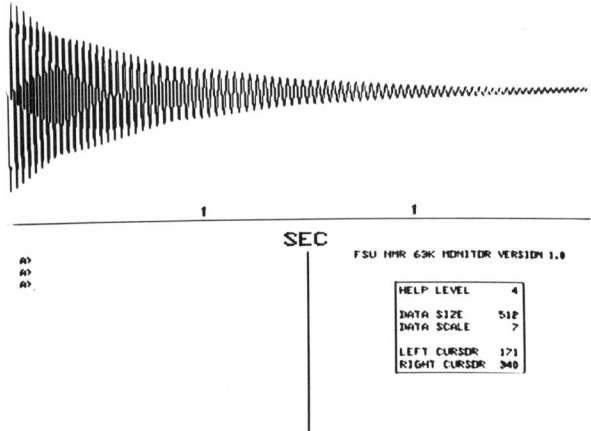

FIGURE 5.18.   *A typical video display of the Conversational Command Interpreter showing the 512 x 256 data area on top, the terminal area on the lower left, and a help area on the lower right.*

can be selected to tailor program prompting to the level of proficiency of each operator. This is a useful feature in a multi-user laboratory where frequent personnel changes ensure a wide spectrum of operator abilities ranging from "novice" to "expert". The features of each of these four levels are outlined below:

LEVEL 1: An expert operator often wishes to execute a series of commands as rapidly as possible with no distracting feedback from the computer. Level 1 provides command execution with no frills, resulting in only the command mnemonics displayed as entered by the operator to serve as a record of the most recently executed commands.

LEVEL 2: The most commonly used mode of operation displays the name of each command to the right of the user entered mnemonic. The operator then can see not only the command mnemonic, but also the command that this mnemonic represents.

LEVEL 3: An operator who is still learning the command structures of the program may opt for level 3. In this level, not only is the command name displayed, but up to a paragraph of additional descriptive information about that command is printed in the lower right quadrant of the screen.

LEVEL 4: This level serves as the default execution level. It provides the greatest amount of interaction for novice users.

In addition to all of the features mentioned previously, level 4 offers a conditional execution interlock. The command name and description is displayed as before, but execution does not begin until the operator gives his consent. In this way operators can become familiar with new command mnemonics without fear of unwittingly or irretrievably modifying their data. An example of Level 4 operation is shown in Figure 5.19 illustrating the Help Level (HL) command that permits modification of the current level of interaction.

FIGURE 5.19.   An illustration of Level 4 command execution. The command name is given on the left and descriptive information is shown on the right. This command will not be executed until the user types a Y. Data zooming for increased resolution is also illustrated in the data area.

Another level of execution is built into every command. Level 0, invisible to the operator, provides a "batch" mode of operation. Some commands do not lend themselves to batch execution, so for these commands level 0 is simply a non-execution level.

The software architecture of the mappable PROM banks and the structure of the command decoder itself have both been carefully designed to make the incorporation of new applications commands as simple and dynamic as possible.

The command decoder accepts any two letter keyboard input and searches all possible banks for a match. If a command mnemonic is found that matches this input, it is executed at the currently defined help level. If no such command is found, the operator is informed that the command does not exist. No restriction is placed on two letter combinations, except that each one must be unique. With such a structure, new commands can be incorporated into the system by simple incorporation into an existing command bank. No modification to the CCI is necessary at all. Because of this dynamic command environment, one command is provided that searches the PROM banks and informs the user of the possible legal commands in a given system.

The first section of any PROM bank is allocated as a command look-up table. Each entry in this table consists of the two letter command mnemonic, followed by an options word describing the command, and the starting address of the command. The normal command table allows room for up to 42 unique commands in a given bank, but this number can be increased if necessary. The terminating value of the look-up table consists of a byte with all bits set to logial one. This is convenient since it is the value read if a bank is physically not present in the system, and it is also the value of an unprogrammed PROM location. If a bank is not in the system, the command decoder simply treats it as having a look-up table of zero length. If a command is to be added to an existing bank, it can be done by making the appropriate entry at the end of the look-up table, and supplying the PROM chip containing the software for that command. From the perspective of the command decoder, the PROM banks are indistinguishable and completely location independent.

To maintain compatability with the command decoder and the help level structures discussed above, each command must have a minimal amount of common internal structure. For ease of programming, this common structure is placed at the beginning of the command. It consists of two entry addresses, the first for level 1 execution and the next for batch (level 0) execution. Following these addresses is the command name as printed in level 2, and the descriptive information used in level 3. Such a structure allows each command to be completely self-contained and makes it easy to adapt new software to the constraints of the CCI.

As mentioned earlier, 48K bytes of RAM memory have been allocated for spectral data sets. However, this represents the maximum length of a data set. The majority of real data sets are substantially shorter than this. In order to take advantage of the excess memory that would not ordinarily be utilized, the CCI supports up to four user-defined logical data areas. Each data area, designated A, B, C or D can be independently configured as 16- or 24-bit words, single or dual channel, and any length, as long as the total of all data areas is less than the memory available. The operator remains in blissful ignorance of the exact location of these data areas, changing logically from one to another with a simple command. All reformatting and memory allocation is done transparent to the operator. In order to maintain a unique identity, each assigned data area carries with it a 1K byte header that contains all of the information necessary to uniquely define its contents, including size, data collection parameters, and a user supplied dataset name and comment. Multiple data areas can prove to be invaluable in experimental set-up, when searching for appropriate nmr pulse widths or when comparing two spectra for differences. Another powerful application of multiple data areas is in "tau-cycling" during a long nmr T_1 measurement. In this example, data accumulation can be carried out sequentially into several data areas, providing an averaging effect both for duty-cycle (hence sample heating) variations and for long term experimental drift.

The four physical memory banks of a basic INDIAN system are logically divided into nmr-specific and non-nmr banks. The first bank contains general support commands used to interact with the video display and the keyboard. These include parameter entry routines and display routines such as a "zoom" (horizontal expansion) function. The data area remapping commands are also located in this bank. Another general purpose bank is allocated to the WARPATH network interface commands, giving the operator the capability to send and receive files on the network. Part of this bank is also used for a general purpose program that allows the development of specialized timer sequences in a high level language especially designed for use with the PTPC timer board discussed earlier in this chapter. These sequences can either be executed immediately by the CCI, or stored via the network as experiments tailored to the needs of a particular user.

One of the nmr banks is dedicated to data processing commands, including the Fourier Transform, exponential multiplication, limited spectral phasing, and other commands necessary for basic FT nmr data processing. All of these processing commands operate on 16-bit data words only, since they are designed to be used only for setting up experiments, a situation where high dynamic range is not important. Another nmr bank is allocated to data acquisition commands, putting the DAS hardware through its paces in a wide variety of specific preprogrammed nmr data acquisition sequences.

<u>INDIAN / WARPATH Interface</u>. INDIAN data stations can be implemented without directly attached hard-copy or mass-storage devices. Versatile links to a network containing such resources are thus vital to the complete and efficient functioning of such stations. The WARPATH local area data network, with its intelligent interfaces, is discussed in detail in an earlier section of this chapter. Since each INDIAN already contains a Z-80 microprocessor -- one element that makes the WARPATH interface intelligent -- a somewhat less intelligent S-100 based WARPATH interface has been devised for the INDIAN.

The S-100 WARPATH interface consists of the differential signal transceivers and Intel 8291 Talker/Listener integrated circuit as in a standard intelligent interface, and also includes some simple S-100 interface logic. It does not contain the microprocessor, memory or host interface logic of an intelligent interface, since these components are assumed to be elsewhere in the S-100 system. The elimination of these components in the S-100 WARPATH interface has both positive and negative implications. The primary negative aspect involves network interface standardization (or lack thereof). Since the S-100 WARPATH interface does not include its own microprocessor, the network software is of necessity somewhat different than that of the other interfaces in the network. These differences, although minor, result in somewhat different data flow, increasing slightly the conceptual complexity of the network. On the positive side, this same software difference allows the elimination of one layer of data buffering. Data being transferred to or from the network can be handled directly by the host Z-80, with no need for separate host software. The elimination of the go-between network processor also increases overall data throughput, since the block I/O of data now goes directly from host memory to network. Possibly the most important positive implication of the simplified S-100 network interface is that __all__ network software at this node must execute on the host processor, allowing access to sophisticated systems level software development and debugging tools when implementing new network software. Routines can be thoroughly tested in the S-100 environment before being committed to an intelligent interface where they become much less accessible for direct debugging.

WARPATH network access is built directly into the structure of the INDIAN software. With this software a user can specify that data can be transmitted to or from any active node in the network. In addition, automated data transfers can be carried out during the course of a long experiment to provide intermediate back-up copies of critical data, or in the case of experiments that generate several data files, to store these files at another node with no user interaction. This provides the user with the capability to begin processing the first few files or to look at intermediate results of an experiment at the processing or storage node of the network while data collection is still underway at the INDIAN node.

THE DATA PROCESSING STATION

The ultimate destination for most data collected by the INDIANS on the WARPATH is a Data General Eclipse^{TM} S/130 minicomputer running the Advanced Operating System (AOS).
The minicomputer and its operating system support two concurrent users running high speed interactive graphics software designed to process large spectral data sets. In addition, other users not using the graphics terminals can simultaneously program or run standard programs on conventional terminals. All users share spooled access to system peripherals, such as fast printers and a high-speed digital plotter. Data security is provided by a system of usernames, passwords and access control lists, designed to protect individual directories and files from unauthorized modification. In short, the processing station provides many of the features found in a mainframe-based computer facility, as well as features uniquely suited to the laboratory environment.

Processing Station Overview: Hardware. The instruction set and architecture of Eclipse computers are supersets of the Data General Nova^{TM} line computers. In addition, the S/130 model supports floating point instructions (firmware) as well as a microprogrammable Writable Control Store. The word length is 16 bits and the cpu can directly address 2^{15} = 32k words of memory (the remaining bit is used for indirect addressing). Clearly, this is insufficient for a moderate-sized multi-user environment. Larger memory configurations (up to one megabyte) are accommodated by a Memory Allocation and Protection (MAP) unit whose function is to translate logical memory addresses to physical addresses.
The Eclipse peripherals are divided into two categories -- standard and intelligent. Intelligent peripherals were purchased or constructed whenever possible in accordance with distributive processing principles. The Z-80 based network interface, for example, relieves the Eclipse of primitive network control responsibilities. The microprocessor facilities in the plotter (based on an Intel 8085) are quite sophisticated, providing a variety of functions such as character and line generation. The videographics generators are also intelligent, with dedicated bit slice microprocessors providing high-speed 512 x 512 point raster scan displays.
The videographics microprocessors can communicate with the host CPU via the data channel (DMA) and consequently, are potentially extremely fast. Unfortunately, they were not designed to be graphics terminals. A little bit of hardware and software manipulation, however, has resulted in an acceptable arrangement as shown in Figure 5.20. This configuration does burden the host computer by requiring a resident manager program

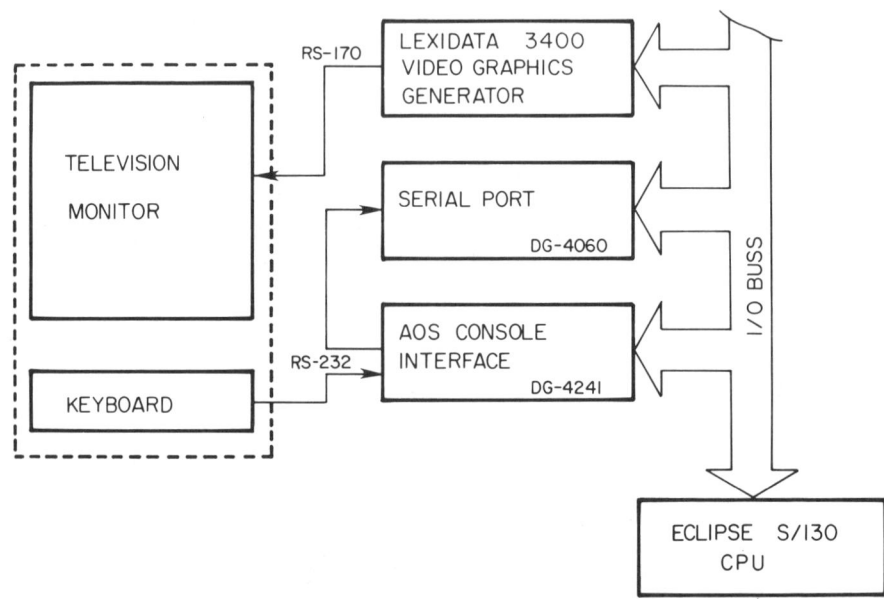

FIGURE 5.20. *Construction of a high speed graphics "terminal" using standard serial interfaces, a keyboard and a high speed (DMA) video graphics generator.*

to service the serial I/O port, but any loss in efficiency is more than offset by the gain in speed of this graphics "terminal" over typical serial-interfaced graphic terminals.

Processing Station Overview: Software. The processing station is in some respects a small computing center. The computer uses a sophisticated operating system (AOS) that can provide all the services of a large main frame facility (but on a much smaller scale). A systems console is used to monitor and control up to 64 memory independent programs (or processes). Users can start processes on other consoles only after providing a username and password to the operating system. Users have access to a wide range of capabilities ranging from automatic print and plot spooling to delayed batch execution of programs. The spooling feature permits multiple users to send output to a peripheral even if the device is busy with another output file. Furthermore, the user does not wait for completion of the output, but regains control of his program almost instantly after submitting an output file to the spooler. Each user also has a personal directory protected from all other users. The "big computer" structure provides a secure, easily learned architecture, flexible enough to meet diverse laboratory needs.

The runtime environment of the computer consists of several independent programs running concurrently. Some programs such as the printer spooler and peripheral manager are part of the operating system while others (the plot spooler and graphics display manager) were written in-house. All programs fall into one of the three categories shown in Figure 5.21. Most of the

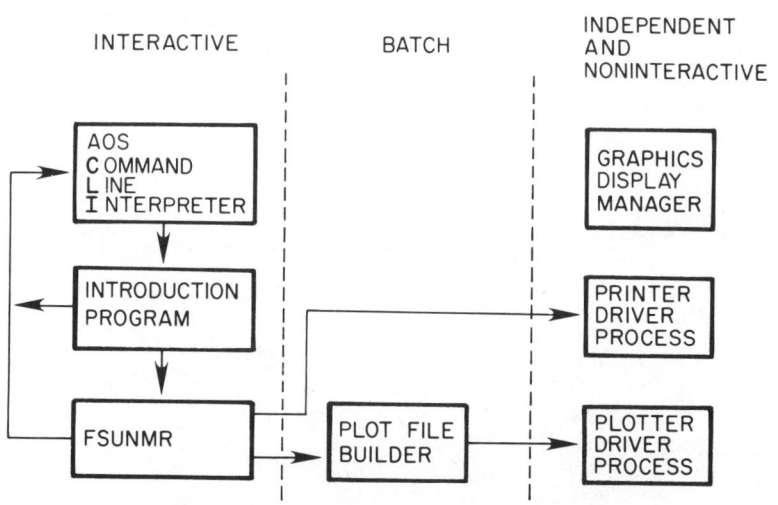

ECLIPSE PROCESS ENVIRONMENT

FIGURE 5.21.   Three types of programs in the minicomputer environment. Arrows indicate the activation of a program by another.

processes require user interaction and therefore are associated with some sort of terminal. Conversely, some programs are designed to be sequentially executed in the batch stream without a terminal. All other processes are initiated by the operator at system startup and manage various peripheral devices. The casual user directly interacts only with his program and is generally unaware of other processes being executed by the operating system and the laboratory software package.

The primary function of the Eclipse as an interactive data processing center is accomplished by two types of programs:

utilities and the main program, FSUNMR. Three utility programs run concurrently with operating system utilities, terminating only when the operating system terminates. The graphics display manager (DSMGR) is a resident program that emulates terminal functions for the graphic displays. Figure 5.20 shows how all alphanumeric characters to be displayed in a normal terminal mode are presented to the serial port. Notice that these characters come from the console interface regardless of their origin--keyboard echo, system response or program response. DSMGR monitors two serial ports, one for each display, and plots the incoming characters at the appropriate location in the alphanumeric field. Normal graphics are directly under program control, bypassing DSMGR, and are thus operating at DMA speeds.

Plotting occurs via a complex mechanism illustrated in Figure 5.22 and described on the following page that utilizes the plot manager utility (PLMGR).

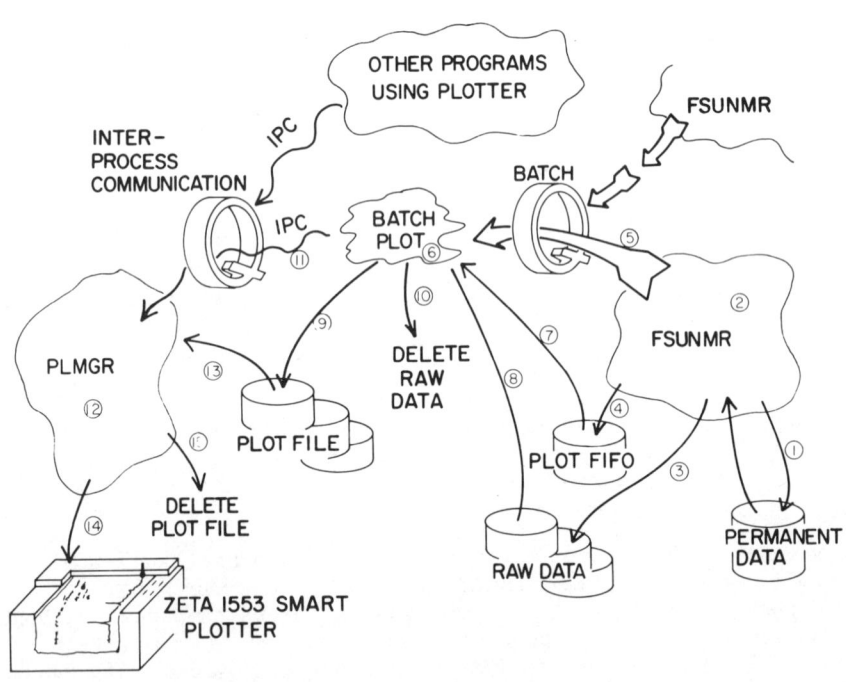

FIGURE 5.22.   *The plot enqueing mechanism constructed to spool plot outputs. A brief description of each numbered step is included at the right.*

Plot Enqueuing Mechanism

1. Data sorted in the user's directory is called into the main program.

2. The user then processes the data and enters the "PL" routine when ready.

3. "PL" generates an unused file name, creates a file by that name, and writes the data to it.

4. The file name is appended to a file called "PLOTFIFO".

5. A batch job is initiated.

6. If the batch queue is empty, "BATCHPLOT" will run, if not, "BATCHPLOT" will wait to be run. Note that every plot initiates a "BATCHPLOT".

7. "BATCHPLOT" gets the first file name off "PLOTFIFO".

8. "BATCHPLOT" loads that data set into its memory area.

9. "BATCHPLOT" converts the raw data set to plotter format and writes it back to disk.

10. The raw data file is deleted.

11. "BATCHPLOT" enqueues an "Inter-Process Communication" (IPC) to the plotter manager and stops. The batch queue is now ready for the next "BATCHPLOT".

12. "PLMGR" activates itself upon receiving the IPC.

13. "PLMGR" loads the plot data into its memory space.

14. "PLMGR" transfers the data to the plotter one block at a time. The plotter checks for transmission errors at the end of each block. Should an error occur, the block is retransmitted.

15. "PLMGR" deletes the plot data file.

After initialization, PLMGR suspends itself and waits for an "Inter-Process Communication" (IPC) from a process that has prepared a file for plotting. The complete name of the plot file is passed in the IPC so that PLMGR can open that file and transmit it to the plotter. The file is sent in blocks terminated with a checksum. Any transmission errors are detected by the plotter which initiates a retransmittal of the bad block. The operating system could have easily been made to provide the plot spooling function, but error checking would have been impossible.

The WARPATH service routine (WPMGR) is still in the design stages. The overall mechanism has been formulated, however, and is shown in Figure 5.23. This arrangement allows data transfers to be initiated from any point in the network, or from the Eclipse. Thus, a user at a data acquisition station can place data into his Eclipse directory without logging onto the Eclipse. Likewise, a user logged onto the Eclipse can retrieve or send his data to or from any currently available network station.

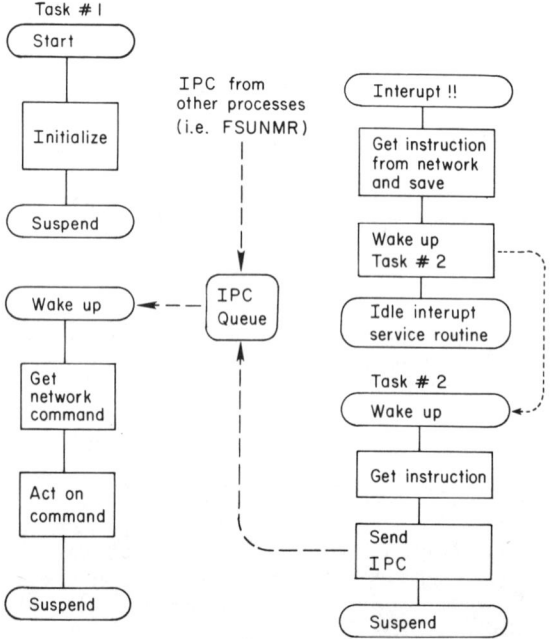

FIGURE 5.23.   Flowchart of WPMGR, the WARPATH interface software. This is the only multi-tasked program. A second task is necessary because interrupt service routines are not allowed to send IPC's in this operating system.

In addition to the three user-independent utilities (DSMGR, PLMGR, and WPMGR), there are two user-dependent utilities. Both are written in FORTRAN 5[5] and run only when necessary. The first is an introductory program executed prior to FSUNMR. This program offers the user helpful information about the program he is about to enter. Descriptions about commands, screen and data formats are presented upon request. When the user is ready to enter the main program, the introduction utility terminates and FSUNMR is automatically started.

The other-user dependent utility runs in batch mode and is part of the data plotting sequence described in Figure 5.22.

All of the utilities support the main program, FSUNMR. FSUNMR is a large program currently using about 40 overlays and a virtually addressed data space constructed for 32,768 floating point numbers. The program is highly modular, consisting primarily of subroutines written in FORTRAN 5. (Some modules have been written in assembly language to increase execution speeds, but these modules are quite small and simple.)

For the most part, FSUNMR is written to be machine-independent. Since the program does rely heavily on the graphics displays, some routines are specific for the hardware configuration. The number of these routines, however, has been kept to a minimum to simplify implementation of this software on other computers.

Large data arrays are possible because of an assembly level virtual array subroutine written in-house that is both machine- and operating system-dependent. The FORTRAN code, however, can be easily modified so that array references are executed normally. Consequently, implementation of these arrays on other machines supporting FORTRAN should cause no problem. Whenever a memory reference is made to a location in the virtual array, the index of that location is passed to the subroutine. Two memory windows are managed for each array as shown in Figure 5.24. If the desired location is not within one of these windows, the least recently addressed window is remapped by a system call to another page in the computer's memory. The operating system treats the virtual array as a file, keeping as many pages as possible in memory, swapping pages to disks heuristically and transparently. The data is then retrieved from (or stored to) the appropriate location in the window area. Two window areas are more often efficient than one because some algorithms perform multiple memory references in an alternating fashion. A trivial example is given below:

```
      DO 100 I = 1, MAXINDEX/2
      K = MAXINDEX-(I-1)
      TEMP = DATA (I)
      DATA (I) = DATA (K)
      DATA (K) = TEMP
  100 CONTINUE
```
The virtual array subroutine does keep the FORTRAN code machine-independent, but at a high cost. Memory access times are typically an order of magnitude slower than normal addressing.

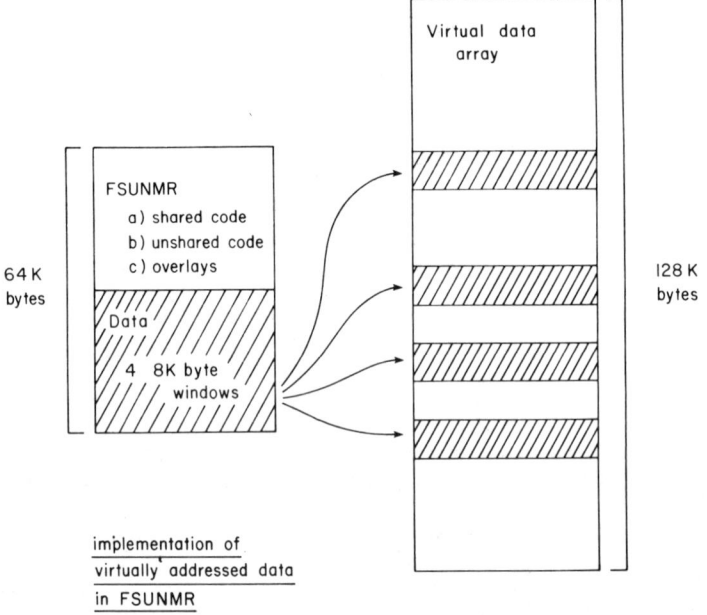

FIGURE 5.24.   Memory configuration of FSUNMR. Half of FSUNMR consists of the main data table windows. There are two main data arrays, each with two windows. Each array is 16,384 floating point numbers long (64K bytes).

Another mode in which remapping operations are performed explicitly in the FORTRAN code permits direct addressing of those locations within the window. The code for such routines is very non-standard, however, and thus this technique is reserved only for those routines where speed is critical (The Fourier transform and display routines).

Design: The choice of an operating system is critical in the design of any multi-user facility. A broad range of operating systems are usually available for any given computer. Minicomputer operating systems can be quite small and simple, supporting no more than two programs at a given instant in a background/foreground fashion; or they can be quite complex, simultaneously supporting a large number of programs. Quite often small operating systems execute multiple program execution paths within the same logical memory address space. Many computers have hardware devices that enable segments of main memory to be treated independently of each other, thus allowing programs to be totally memory independent. Multiprogramming operating systems that make use of memory remapping provide a greater degree of flexibility and security than background/foreground operating systems. In addition, they are usually powerful enough to provide large-computer features desirable in a multi-user facility. A multiprogramming operating system was chosen for this laboratory data center primarily to provide both users and programmers with the highest degree of flexibility.

The task of designing, writing and maintaining software in our laboratory has been minimized by employing certain techniques and philosophies[6,7]. First, the overall structure as well as individual routines are designed in a top down manner[8]. No code is written until the software is completely designed. Major software interfaces are built and tested long before specific routines become available.

The code is written in discrete, independent modules to facilitate debugging and modification. Modules are constructed to be monofunctional and general purpose. Modules constructed in this manner can be treated as interchangeable "black boxes" whenever necessary. The Hierarchical Input Processing Output (HIPO) diagram shown in Figure 5.25 indicates how modules can be connected to form large programs such as the one used in our laboratory. Finally, a consistent programming structure (or style) is enforced whenever software is written. This technique turns out to be particularly important since several programmers often contribute to a given project or module during all phases of its existence. Thus, any programmer familiar with the structure can easily repair or modify a module even if he did not write it. Furthermore, modules can be written more quickly if a structured approach is used.

It should be noted, however, that these techniques occasionally result in less efficient code at run time, so some critical routines have had to be designed without them. In general, however, a great amount of effort has been saved and channeled elsewhere as a result of adherence to these guidelines.

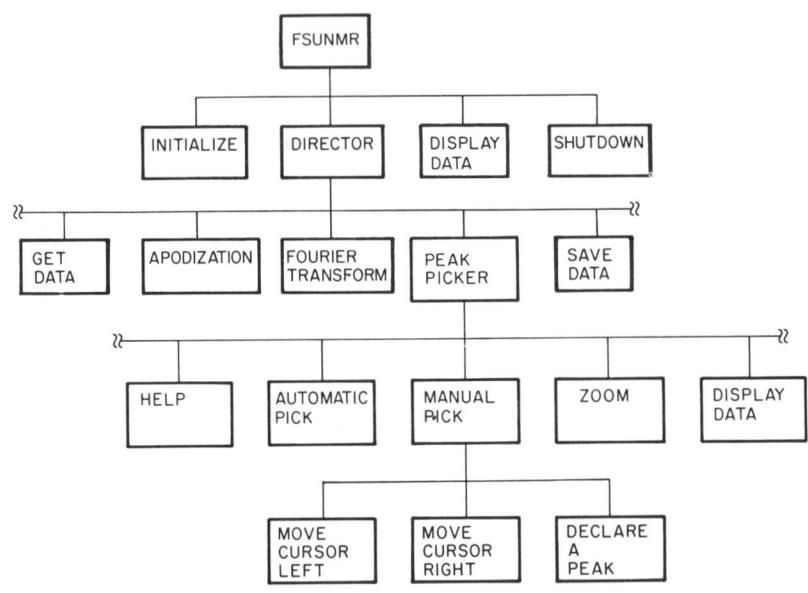

FIGURE 5.25. *Partial Hierarchical Input Processing Output (HIPO) diagram of FSUNMR highlighting the peak picker. A HIPO chart such as this one graphically illustrates the degree of modularity of FSUNMR.*

Since the software under development in our laboratory is to be used primarily by non-programmers, it has been written with the user in mind. Situations in which users can crash or hang programs are rigorously avoided. Input is generally checked by the software and rejected if inappropriate. All iterative loops are permitted a finite number of iterations to prevent program hangs. In addition, users are always made aware of their options through displayed option lists and extensive help sequences (there is even a key marked "help" on the keyboard to aid the inexperienced user). The main program, FSUNMR, offers the user an extensive collection of processing routines, some providing identical functions but employing different algorithms. This sort of redundancy enhances the program's utility and power since the user is frequently the best judge of which algorithm to use for a given data set.

FSUNMR and the User. Once the data is acquired at an INDIAN and sent to the data processing center, the user is ready to process his data. The user begins a data analysis session by logging onto the Eclipse at one of the graphics consoles. Infrequent users who do not need the programming features of the operating system are forced into the Introduction utility. Other users can enter the operating system's Command Line Interpreter (CLI) and then, if desired, enter the FSUNMR package.

The Introduction program offers the user a list of options, the first of which is to go to FSUNMR. The other options are intended to prepare the user for the main program, with text and diagrams relating to several subjects, such as command structure, data representation and display format.

Upon entry into the main program, the graphics screen is erased and the various display areas are initialized. All keyboard echoes and formatted program responses appear in the lower left quadrant of the screen under the control of DSMGR. Option lists and help information appear in the lower right quadrant and are not under the control of DSMGR. These messages are short disk files that are displayed by a routine interacting directly with the graphics display. Spectral data is displayed on the top half of the screen with a resolution of 256 x 512 picture elements. Since most data sets consist of far more than 512 points, a powerful "zooming" routine is available to expand selected display regions so that the display resolution can exceed the number of data points. A small strip immediately below the data region is utilized for scale information and units (typical display, Figure 5.26). Together, the four display areas provide the user with a convenient and consistent visual format throughout the program (the display format serves as a model to the one used in the INDIAN graphics terminals).

FSUNMR is a command-oriented program. Main commands consist of two characters. Execution of a main command starts immediately after entry of the second character. Several main commands are extended with subcommand structures based on one character subcommands. Whenever a subcommand is expected, a modified prompt is displayed in the terminal field (lower left display quadrant) with a current options list presented in the options field (lower right quadrant). Approximately 80 main commands are currently implemented, 15 of which have subcommand structure. New commands can be readily added with a minimum of program modification.

All extended commands have "help" as one of the options. In addition, an entire extended command is dedicated to aiding the user. In principle, the user has all the information necessary to run the program at his fingertips without need to consult a manual (although one is available).

All printing and plotting is spooled. In other words, the user and his program are not burdened with details concerning output. Spooling has two important ramifications: (1) the user regains control of the program (and thus, the ability to continue processing) immediately after his request and before the print or plot appears (in fact, multiple prints and plots may be initiated for one or more data sets); and (2) multiple users share the same device and requests are serviced on a First-In First-Out (FIFO) basis. The print spooler is a part of the Data General operating system but the plot spooler, the mechanism of which is shown in Figure 5.22 was built in-house.

Sample Capabilities. The heart of any FT spectroscopic analysis system is the FT routine. Most modern applications use a Fast Fourier Transform algorithm. While much faster than the older, Discrete Fourier Transform, FFT routines still tend to be one of the slowest parts of a processing sequence. The FFT routine of FSUNMR[9] is written entirely in FORTRAN 5 except for the bit reversal routine which is written in assembly. The routine is written for floating point data and consequently the data do not require scaling during the computation (unlike fixed point FFT algorithms). Assembly FFT algorithms generally use a sine-cosine lookup table to minimize execution times. We have found that our FFT subroutine is actually faster when sine and cosine functions are used in place of a FORTRAN lookup table. The FFT algorithm currently in use in FSUNMR can do a 8192 point transform in 8 seconds when the data are addressed directly. The algorithm can become very inefficient, however, when data are addressed virtually. Our FFT algorithm minimizes the amount of remapping by explicitly remapping memory. Consequently, its speed approaches that of a directly addressed algorithm for large data sets.

FSUNMR incorporates all the standard features of any Fourier Transform software package. The manner of operation, however, is unique. Many complex functions can be accomplished automatically, but several levels of manual operation remain optionally available for the user. The Peak Picker (PP)[10], for example, automatically calculates the noise level of a spectrum, identifies peaks above this threshold and calculates line position, shape, and total peak areas. The peaks are then labeled on the screen and stored in a table. If the selection was unsatisfactory, the user can use subcommands displayed in the Peak Picker options list to employ any of several manual techniques to modify the table so that only desired peaks are present. The user can delete or insert peaks in the table, or he can change the threshold and allow the automatic peak picking algorithm to try again. The peak picker is a prerequisite to some commands that require a peak table; therefore, several of the displayed options provide automatic entry to other routines.

<u>C</u>alculate <u>R</u>elaxation (CR) is a good example of a routine requiring the peak table. CR uses the peak position listed in the table to identify resonances in a series of spectra acquired in spin relaxation multiple pulse NMR experiments. The T_1 of each peak is then calculated with an iterative three-parameter exponential fit[11] and the results presented to the user. At this point individual exponential curves can be displayed or plotted by the user, but in CR the data values themselves cannot be manipulated. Manipulation of this relaxation data is available, however, in another routine caled <u>R</u>elaxation <u>A</u>nalysis (RA).

In RA the user identifies the peak of interest (via the peak table or, optionally, with displayed cursors) and uses the three parameter fit to calculate a result. The resultant curve is displayed, along with a statistical analysis for each data point. The user can ignore, delete, change or insert individual data points. Several methods of data plots are available for final output, along with hard copy of the statistical analyses.

Different levels of control over an algorithm are also evident in the <u>B</u>aseline <u>F</u>lattening (BF) routine. BF is a powerful algorithm based on one by Pearson[12] which is designed to remove broad baseline distortions, leaving narrower spectral features, such as peaks (and their integrals) and noise, intact. The user has control over the parameters used by BF and can choose, for example, the number of independent regions to be flattened. Figure 5.26 shows direct photographs of the graphics displays of a badly distorted spectrum before and after baseline flattening. Note that even small real lines are unaffected by the flattening algorithm, provided only that those lines are narrow.

Since most experimental measurements are subject to baseline distortions to some extent, baseline flattening is an extremely useful tool if quantitative integrations on discrete data are desired. Long NMR experiments, for example, can produce data with large distortions. Even small distortions such as a small offset in the data can introduce significant errors in the integration of that data. If the distortion of a baseline is removed, however, numerical integration techniques become meaningful. Figure 5.26 illustrates the integration of a data set with and without Baseline Flattening. Note that without Baseline Flattening a small offset in the data caused substantial errors in both the displayed integrals and those calculated by Simpson's Rule. Baseline Flattening prior to integration, however, removes the distortion permitting quantitative integrations. (This procedure is far more valid than the customary DC and ramp offsets applied by user-selected knob adjustments of nmr data).

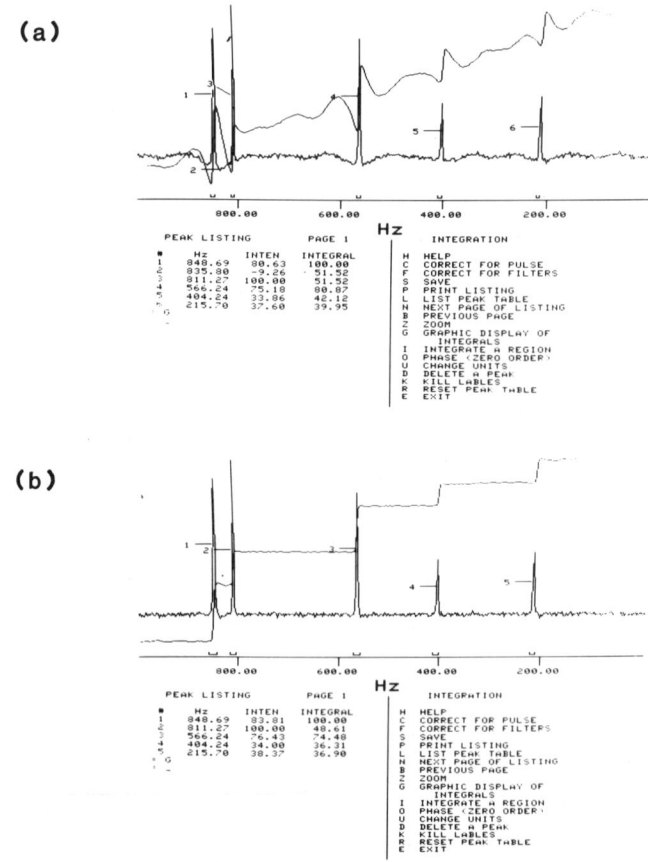

FIGURE 5.26. (a) The videographics display showing a deliberately distorted spectrum after automatic peak picking but before baseline flattening.
(b) Display of the same data after baseline flattening. Note dramatic improvement in peak integration curve.

FSUNMR and the WARPATH Environment. The FSUNMR software system was designed to provide optimized data processing in the multi-user laboratory environment. This is accomplished through a combination of powerful high level algorithms and a multi-layered control hierarchy supporting automatic, interactive, and hybrid modes of operation. The user operates in a humanized environment implemented with fast graphics I/O. The design philosophy inherent in FSUNMR is also implemented in the WARPATH computer network and the network data acquisition stations. For example, network operation is largely transparent to the user, with network status displayed on graphics monitors. INDIAN microcomputer data acquisition software strongly resembles FSUNMR from the user's point of view. INDIAN hardware also utilizes fast raster scan graphics I/O.

WARPATH, FSUNMR, and INDIAN (hardware and software) are similar in that highly modular designs are employed. This facilitates maintenance or modification of these subsystems. These three subsystems form a synergistic laboratory data system that increases experimental data throughput and provides unique data processing capabilities.

CONCLUSIONS AND ACKNOWLEDGEMENTS

In no scientific field are new developments coming as rapidly as in computer networking. Tomorrow's implementations of Ethernet[TM] and similar contention networks will make obsolete most existing schemes. Network designs such as WARPATH serve as training grounds, easing the transition from dedicated mini- or micro-computers to extensive and powerful laboratory computer networks--the configurations of the late 1980s.

Our original entry into computer networking was prompted by Dr. David Wright, now on the Chemistry Department staff at the University of Illinois. David recognized very early the advantages of the distributed approach for laboratory data acquisition and processing. Following his germinal work on WARPATH Dave Wright has given invaluable assistance in the design and development of several aspects of the DAS hardware.

Our Z-80 based data acquisition hardware utilizes concepts from the system developed by Charles N. Reilley (Chapter 2). We would like to acknowledge important hardware design contributions from Jaan Past (an exchange visitor from Endel Lippmaa's laboratory in Estonia), Professor Robert Craig (University of Melbourne), and Richard C. Rosanske and Dave Roberts (both at Florida State University). Software design contributions from Professor F. A. L. Anet (UCLA), James Dumoulin and Renee Roeder (M.S. and B.S. students in computer science, Florida State University), and Jackson Read, G. William Coppenger and Wayne Joubert have been extremely helpful.

Finally, we wish to acknowledge financial support from the National Science Foundation for aspects of this research.

Final Note: This chapter was produced using WORDSTAR[TM] text processing software on the Z-80 microcomputer development system.

REFERENCES

1. J. W. Cooper, *Topics in Carbon-13*, 2, Chapter 7.

2. R. Kaiser and W. R. Knight, *Journal of Magnetic Resonance*, 36, 215 (1979).

3. T. A. Case and H. T. Stokes, *Journal of Magnetic Resonance*, 35, 439 (1979) and references therein.

4. M. Dahmke, *Byte*, 5, 196 (1980).

5. FORTRAN 5 is a Data General extension of ANSI FORTRAN IV. The compiler globally optimizes code in 11 passes, producing highly efficient executable code.

6. Fraser, A. G. (1969) in *Software Engineering*, Naur, P. and Randell, B. Eds., NATO Scientific Affairs Division, Brussels 39, Belgium.

7. Van den Bout, G. A. *Byte*, 4, 176 (1977).

8. Yourdon, E. (1975) in *Techniques of Program Structure and Design*, Englewood Cliffs, N. J., Prentice-Hall.

9. Dumoulin, C. L., Levy, G. C., Anet, F. A. L., *Computers and Chemistry*, submitted.

10. Two letter command mnemonics for FSUNMR have been selected to follow commercially available software (Nicolet, Inc.) when similar functions are utilized.

11. Kowalewski, J., Levy, G. C., Johnson, L. F. and Palmer, L. (1977), *Journal of Magnetic Resonance*, 26, 553.

12. Pearson, G. A. (1977), *Journal of Magnetic Resonance*, 27, 265.

6

HARDWARE AND SOFTWARE ASPECTS OF THE DEVELOPMENT OF THE SATELLITE DATA PROCESSING NETWORK AT THE UNIVERSITY OF HAMBURG

K. Zietlow, H. Ch. Broecker, H. G. W. Müller

INTRODUCTION

In the last years the potencies of analytical chemistry have been influenced considerably by the progress of computer technology. In striving for laboratory automation with the help of computers, different conceptions have been realized.[1,2,3] In contrast to the often used stand alone mini and micro computers the hardware and software of which is especially developed for the connected instrument and which control, acquire and process data, in this chapter a satellite computer network is described.

In 1972 several research groups of the department of chemistry at the University of Hamburg began to realize the idea of connecting large analytical instruments to a computer system. This should optimize the faculties of these instruments. On one hand, large data sets had to be correlated to get more information from already existing analyses, and these results had to be presented in a way more suitable to human senses by means of elaborate hardware and software. On the other hand, measurements which involved a great number of repetitive operations had to be automatized. All this done by a computer system would free the chemists from routine work and would increase sample throughput.

For the already used mass spectrometers and gamma-ray spectrometers, which had been the main instruments for this project of computerization, we could buy approved dedicated computer systems with software for data acquisition and data processing. This principle of the dedicated computer had the advantage that the widely differing requirements, depending on the analytical method, of data processing in chemical analysis could be met with in an individual and rather simple but also an expensive way in 1972. A disadvantage, which still exists today, is the lack of intercommunication between such isolated systems.

Our design in 1972 for a computer system therefore contained several small mini computers connected to a larger central computer, so that they could interact.[4] Such a scheme allows small systems with just sufficient hardware to accomplish the specific task with the connected analytical instrument. This means data acquisition at high data rates and real time control of the experiments. After data reduction to minimize the data rate, this small computer can transfer its data to the larger central computer that stores it on a disk for a long time. Sophisticated data manipulation and library search can be done at the central computer without disturbing the real time activities of the satellite computers. The expensive peripheral devices such as rotating memory and plotter for data output can be used by all connected satellite computers.

Such a design, open-ended in a certain way, is more adaptable to the changing requirements of analytical chemistry at a university. Because of these reasons we decided in 1972 to develop a satellite computer network.

HARDWARE OF THE SATELLITE COMPUTER NETWORK

Table 1 shows a list of instruments now connected to the computer system. They are used in five different buildings of the department of chemistry. Most of these instruments are mass spectrometers in connection with gaschromatographs and gamma-ray spectrometers. Their data rates lie in a range from 100 words/s up to 120 K words/s. These are maximum rates, the mean values are more or less lower with the exception of the UV-spectrometer. Its data rate is constant during the whole measuring time. All these instruments are connected to seven satellite computers. If a satellite computer controls two or more instruments, like e.g. computer no. 3, they can only be used alternatively. The parts of the computer network are spread over a rather large area with cable lengths up to 500 m. Fig. 1 gives a schematic outline of the location of the satellite computers and the central computer in the various buildings of the department of chemistry. The small squares indicate the satellite computers and the central computer and the broken lines are the communication lines between them.

The entire hardware configuration can be seen in Fig. 2. The standard type of a satellite computer is the 620/L-100 with 12 K core memory from Sperry Univac formerly Varian Data Machines. These computers have 16 bit word length and a cycle time of 960 nsec. A teletype and a Tektronix 611 storage display are used for operator communication. The satellite computers no. 1 and no. 2 only used for mass spectroscopy have this standard configuration. The hardware interface is similar to the well known one from the "Spectro Systems" of Varian MAT (Bremen).

TABLE 6.1 LIST OF ANALYTICAL INSTRUMENTS CONNECTED TO THE SATELLITE COMPUTER NETWORK

INSTITUTE	ANALYTICAL INSTRUMENT	DATA RATE	SATELLITE COMPUTER
Organic Chemistry	GC-MS Spectrometer 311 A (Varian MAT)	12.5 K words/s	1
	GC-MS Spectrometer CH 7 (Varian MAT)	12.5 K words/s	2
Applied Chemistry	GC-MS Spectrometer CH 7 (Varian MAT)	12.5 K words/s	3
	UV Spectrometer T 13/3 (HDW)	10.0 K words/s	3
Inorganic Chemistry	3 Multichannel Analyzers 8100 (Canberra)	600 words/s	4
Physical Chemistry	Multichannel Analyzer ND 2200 (Nuclear Data)	6.0 K words/s	5
	2 Pulse Height Analyzers 8060 (Canberra)	120.0 K words/s	5
	Rabbit System Neutron Generator	10 words/s	5
	FT-IR Spectrometer FTS 14 (Digilab)	1.0 K words/s	6
Pharmaceutical Chemistry	Tabletting Machine Multiplexer, 13 bit ADC	10.0 K words/s	7

164   The University of Hamburg Date Network

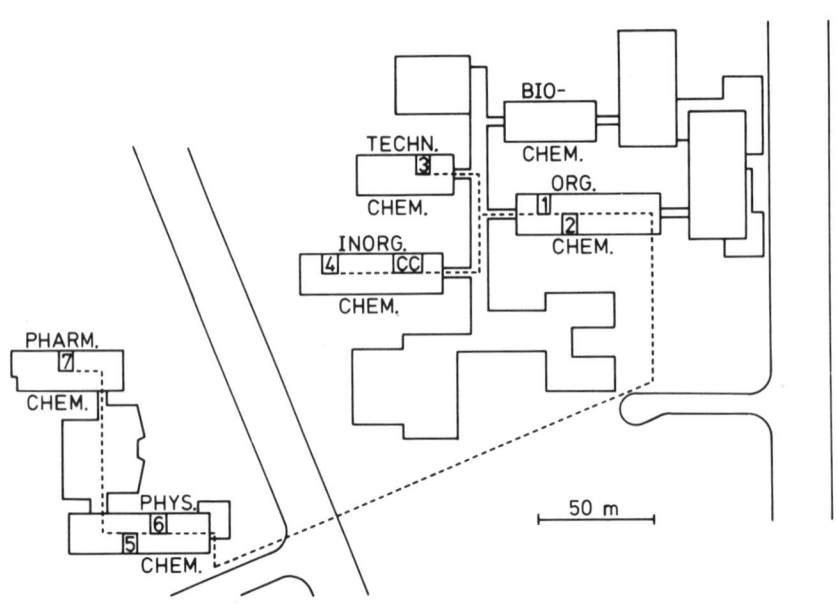

FIGURE 6.1   Schematic outline of the data processing network.

Satellite computer no. 4 is a V71 from Sperry Univac with 32 K memory. The floppy disk is used for data storage. The three multichannel analyzers, parts of gamma-ray spectrometers, are connected via an IEC-bus to this system. A small plotter for output of gamma-ray plots is also available.

The interface for the multichannel analyzer connected to satellite computer no. 5 enables a fast bit-parallel data transfer. Two more pulse height analyzers are connected directly to this computer. Also used for gamma-ray spectroscopy, they are part of a facility for neutron activation analysis with 14 MeV neutrons.[5] The computer controls a neutron generator and sample transport through a rabbit system by switches, valves for the compressed air and IR-light-barriers. It is a fully automatized analyzing system with control of irradiation time, acquisition time and delays. In all gamma-ray spectrometers a double word format is used for the acquired data.

Satellite computer no.6, which is a Fourier transform IR-spectrometer, is a dedicated computer system NOVA 1200 from Data General with a small core memory, a fixed head disk, a plotter and a teletype. Connection to the data processing network has been made to use for the present only the mass storage of

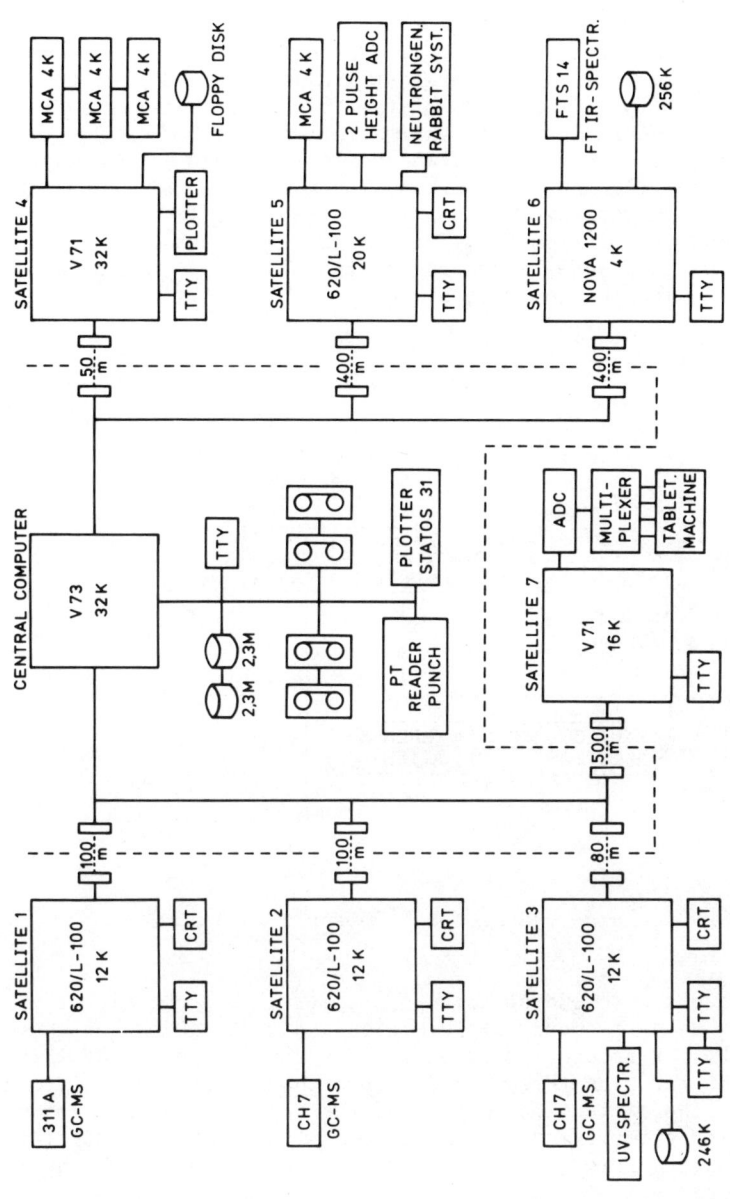

FIGURE 6.2 Configuration of the computer system.

166   The University of Hamburg Data Network

the central computer.

Connection of satellite computer no. 7, a V71 with 16 K memory to the network will be set up this year. A reciprocating tabletting machine with several strain gauges is linked to a standard multiplexer and ADC of the computer. This system is used by a research group in pharmaceutical technology.[6]

Satellite computer no.3, which acquires data from a mass spectrometer too, has an additional fixed head disk because of the high data rate over rather long periods of time of the rapid-scan UV-spectrometer. This instrument scans 100 spectra per second in a wave-length region from 280 nm to 800 nm with an adjustable length from 0 to 300 nm.

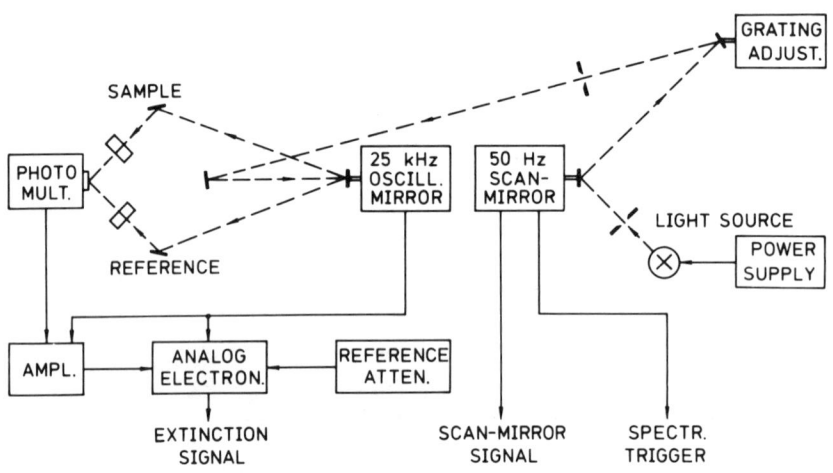

FIGURE 6.3   Rapid-scan UV-spectrometer T 13/3 HDW.

Fig. 3 shows a scheme of the spectrometer. The scan mirror is swinging at a frequency of 50 cycles and makes 100 scans there and back. The second mirror oscillates at a frequency of 25 KHz and leads the beam through the sample channel and the reference channel. The extinction signal and the scan mirror signal from the analog electronics are digitized with two 8 bit ADCs (Fig.4). The two bytes form a data word, which is transferred through opto couplers to the computer. Each spectrum consists of 100 data points, this leads to a data rate of 10 K words/s. The spectra are stored on the disk with a capacity of more than

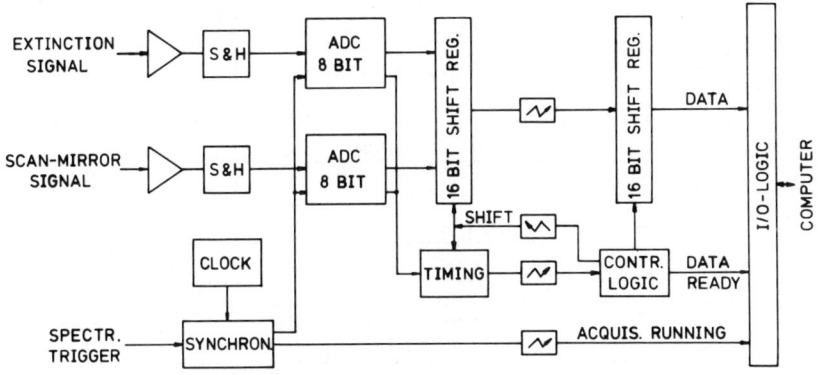

FIGURE 6.4   Computer interface, UV-spectrometer.

2000 spectra. This spectrometer is used to analyze fast chemical reactions.[7]

The central computer is a V73 from Sperry Univac with a rather small core memory of 32 K. It is microprogrammable with 16 bit word length. Four magnetic tape units and two moving head disk drives are used as mass storages. An electrostatic plotter for graghic output and a fast paper tape system are also available.

COMMUNICATION LINES

The communication lines between the satellites and the central computer have been newly developed (Fig. 5). Data are transferred bit serial with a separate clock line at a rate of 30 K words/s (16 bit + parity bit). The clock frequency is about 600 KHz. Three additional control lines are used for the communication link protocol. One signal line initializes and synchronizes data transfer. A second line indicates the CPU step- or run-mode of the remote computer, and the third line indicates an input error from the serial input electronics of the remote computer. Data transfer is fully duplex. The signals are transmitted by differential drivers through twisted pairs and received via opto couplers because of electrical isolation.

168   The University of Hamburg Data Network

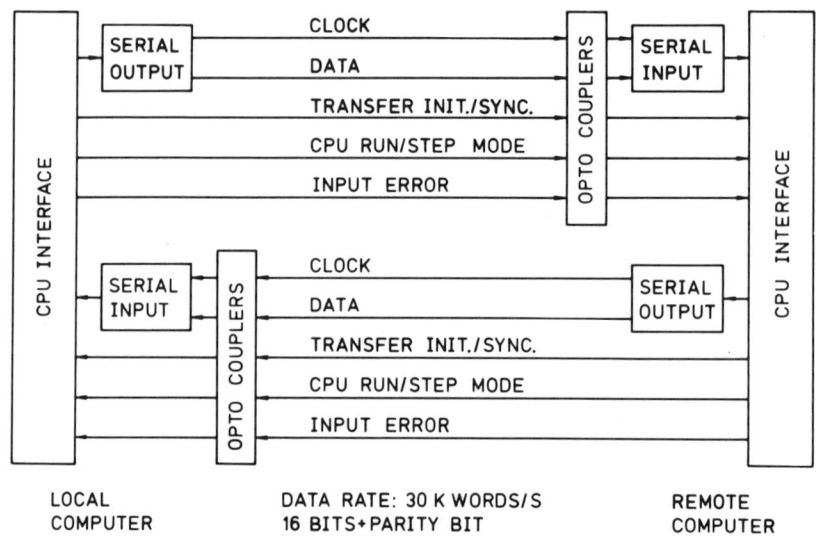

FIGURE 6.5   Block diagram of the communication line.

The opto couplers we used in the beginning got weak after continuously running for about one year. We had an increasing number of parity errors. Now we use opto couplers from Hewlett Packard and have no more such problems.

The high data transfer rate is necessary, e.g. in the case of simultaneous running of the three mass spectrometers. During GC-MS runs with repetitive scans in an average of every 3 - 4 seconds a data set of about 1 K length has to be transferred from every satellite computer. Times for data transfer from the other satellites are not so short. Satellite computer no.6 (NOVA 1200), which is not software compatible to the others, is connected by an asynchronious serial line with a transfer rate of 9600 baud (fully duplex).

COMMUNICATION SOFTWARE AND USER PROGRAMS

Communication Line Protocols.   To the original VORTEX operating system, a real time multi task system, which runs in the central computer, drivers for the communication lines and communication protocol routines had been added. The communication is activated from the satellite computer by sending a signal through the transfer initialisation line to the central computer and generating an interrupt. The protocol routine at the central computer is initiated and responds. A signal exchange follows, a sort of synchronisation. Then the satellite computer transfers a control word, which contains information about the length of an information block to be transferred next to the central computer. This control word is complemented, transferred back to the satellite computer and checked.

The information block transferred subsequently to the central computer contains details about the direction of the following data transfer, the length of the data set, the peripheral devices of the central computer data are transferred to or from. Furthermore the block contains information whether it is loading of a user program or an operating system into the satellite computer, or a parameter transfer to the central computer to initiate there a program with waiting for results. Before starting the real data transfer the central computer transmits a control word for his part, which is echoed by the satellite computer.

At the end of the data transfer there is another exchange of a control word and an end signal through the initialisation line between both computers. By this protocol the effective data rate is slower than the hardware rate of 30 K words/s, depending on the length of the data set.

The operator at the central computer can change the also software controllable priorities of the satellites, he can disconnect a communication line by software, too. Besides these manipulations at the central computer the architecture of the whole system concentrates the initiation of computer processes at the satellite computers.

Operating Systems and User Programs.   The operating system in the satellite computer is called SATKOS (satellite keyboard operating system). This operating system is slightly modified for the different purposes in connection with the controlled instruments. All these systems are in a library located at the central computer. They can be down-line loaded by a simple calling sequence, indicated by the first solid line in Fig. 6, which shows the process structure and data flow of the system. To every operating system belongs a user program library

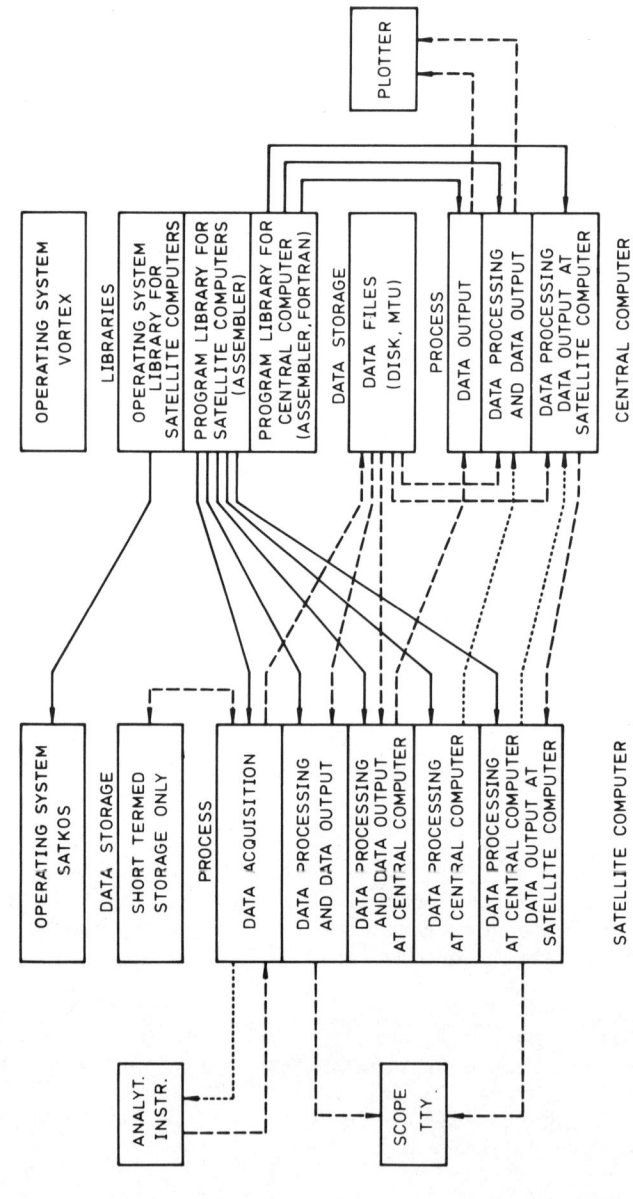

FIGURE 6.6 Simplified software architecture. Detailed explanation will be given in the text.

also in a file at the central computer. The operator can load
and execute such a program by typing a six alpha long mnemonic
on the keyboard. Only assembler programs can be loaded in
machine code with fixed address format into the satellite
computers.
   For data acquisition the programs automatically carry out
data transfer to a previously determined data file at the central computer. This is shown by the broken lines in Fig. 6.
Those data files are made safe by key words against access from
other satellite computers, and against erasure by other operators. Each data set in such a data file or spectrum library
begins with a header, which contains e.g. an operating system
identification, a successive number, a key word, the operators
name, the date, and the name of the experiment. These characteristics can be used by search programs for a detailed data
search later on.
   Data processing and data output can be done in several
ways either at the satellite computer or at the central computer.
The simplified software architecture in Fig. 6 shows these
processes:
1. Data processing and data output at the satellite
   computer.
2. Data processing at the satellite computer and data
   output on the plotter of the central computer.
3. Data processing and data output at the central
   computer.
4. Data processing at the central computer but data
   output on the storage scope of the satellite computer.
The dotted lines in Fig. 6 indicate a parameter transfer.
   Program-to-program communication allows one program in the
satellite and one program in the central computer to execute
and exchange data or parameters in a coordinated manner. The
master program in the satellite computer initiates a slave
program in the central computer and is always in control. The
slave program merely responds to requests received from the
master program, either to accept or reject them. If necessary,
the link between master and slave may be broken by the master
program. For example, the plot programs for data output on the
plotter are such FORTRAN slave programs. Each satellite computer
may have active master and slave programs quasi simultaneously.
   The operator can add to a command loading and executing
an user program several parameters, if necessary, separated
by commas. In doing this, he, of course, has to know the right
sequence of these parameters. The interpreter in the SATKOS
operating system, however, offers the possibility to effect
an input of these parameters in a dialog version. In this

ALUR,X,X,100,3,6,1,3,0

MS - DATA - ACQUISITION   ALUR   01/11/79

SPEC KEY: WA

    1/WA/LP   01/11/79   TEST   CH7TC   WASUM

D,ALUR

START OF SCAN (A/X)	(X)/(?) :	X
TEXT PRINTOUT (S/X)	(X)/(?) :	X
ABS INTENSITY THRESHOLD	(100)/(?) :	100
REL INTENSITY THRESHOLD	(3)/(?) :	3
SAMPLE/PEAK	(8)/(?) :	6
FIRST SPEC#	(0)/(?) :	1
TIME DELAY (IN 100 MSEC)	(3)/(?) :	3
MIN NUMBER OF PEAKS	(0)/(?) :	0

MS - DATA - ACQUITION   ALUR   01/11/79

SPEC KEY: WA

    1/WA/LP   01/11/79   TEST   CH7TC   WASUM

FIGURE 6.7   Parameter input for the user program ALUR (acquisition of low resolved mass spectra): (top) normal version; (bottom) dialog version. Operator input is underlined.

way the operator gets information about the meaning of a parameter and about standard values for the parameters (Fig. 7).

    Another feature of the interpreter is the chance even for a rather unexperienced operator to program the satellite computer in a certain way. He can write a sequence of commands with loops and store it in a special sequence storage area at the central computer. He can reload and execute it.

    We have done a lot of work to give as much information as possible to the user, because most of them are neither scientists nor computer experts. The information system the operator can ask gives answers from a very general form up to a very detailed information, for example, how to use a special program or how to activate a satellite computer.

INFO, INFO

      INFORMATION UEBER INFORMATIONSMODULE.
ALLGEMEINER AUFRUF FUER EINEN INFORMATIONSMODUL
INFO, P1
      P1 = NAME DES INFORMATIONSMODULS
SPEZIELLE AUFRUFE:
INFO,INFO   INFORMATION UEBER INFORMATIONSMODULE
INFO,MODUL   ERLANGUNG VON MODULINFORMATIONEN
INFO,SYSTEM   ERLANGUNG VON SYSTEMINFORMATIONEN
INFO,LUN   INFORMATIONEN UEBER LOGISCHE NAMEN
              DERJENIGEN PLATTEN- UND MAGNETBAND-
              EINHEITEN, DIE SPEKTRUMSBIBLIOTHEKEN
              AUFNEHMEN KOENNEN
INFO,PLAN   INFORMATIONEN UEBER PLAENE

FIGURE 6.8   *General information on use of the information system of the operating system SATKOS.*

Fig. 8 shows a general information on use of the information system. The user can get information about user programs in a more general form, about operating systems, which are running in the different satellite computers, about logical units, where the user can create data libraries, and about the status of the whole system, for example, the status of the peripheral devices at the central computer.

Every user program contains an identification block the operator can load, where he can get the version number, the version date and a full description of this program with its parameters. Fig. 9 shows the identification block of the program NACTAN (neutron activation analysis), which controls the neutron activation analysis facility with neutron generator and rabbit system.

The operating system of satellite computer no. 3 is quite different from the SATKOS system. It uses the same communication line and the same communication driver and protocol routines, but because of the larger memory this operating system is able to run FORTRAN programs.

If the satellite computer is not used in connection with

IDENT,NACTAN

MODULE IDENTIFICATION FOR SYSTEM   AA   01/11/79

NACTAN   07/07/77   VERS.5   AA

            NEUTRON ACTIVATION ANALYSIS

AUFRUF: NACTAN,P1,P2,P3,P4,P5
```
         P1=SPEC-NO (UNGERADE)
            WENN 0 EINGEGEBEN, AUTOM.
            WEITERZAEHLEN
         P2=SPECTRUM KEY (2 ALPHA)
         P3=ACQUISITION CYCLES
         P4=IRRADIATION CYCLES
         P5=DATA TRANSFER PARAM.
            DT:  DATA TRANSFER
            NT:  NO DATA TRANSFER
```

FIGURE 6.9   *Information block of the user program NACTAN (neutron activation analysis).*

an analytical instrument, other systems can be loaded into the satellite like the BASIC interpreter. This is often done by students who solve small mathematical problems during their practical education by BASIC programs.

Program development for the satellite data processing network and assembling and compiling is done at the central computer with support of the VORTEX system. We are now developing a version of a text edit program which runs in the satellite computer and uses the storage capacity of the central computer. By that, program development will be partly possible from the satellite computer and will free the operator teletype at the central computer for operator manipulation which is necessary sometimes.

EXPERIENCES AND CONCLUSIONS

In the first two years of developing the computer network we had only three satellites (no. 1, no. 2 and no. 5). Each of the computers had a magnetic tape unit, so we could use them as dedicated systems with an operation system similar to the formerly used "Spectro Systems" of Varian MAT. This configuration was a very practical one, as on one hand we could test the communication hardware and software extensively with the three satellite computers, on the other hand the connected instruments (especially the mass spectrometers) could be used with the dedicated computers without any disturbance from the satellite network not being fully in operation. The other satellite computers have been added to the system in the following years.

We soon realized that the primarily used mass storage capacity was too small and we enlarged it by another disk drive and two more magnetic tape units. The configuration seen on Fig. 2 is now in operation since 1977. The use of the network has become more and more intense since then. From this sometimes difficulties arise because the memory of the central computer is too small with only 32 K core. Each satellite occupies a memory block for data transfer and for partly resident programs and these memory areas have to be small because five satellites computers share the whole memory of the central computer. This leads to an increased number of accesses to the disk which lays bare another bottleneck. There is only one controller for the two disk drives where we have data files as well as program libraries. During simultaneous GC-MS runs with three satellite computers these insufficiencies sometimes cause the loss of several data sets (mass spectra), because the central computer sometimes cannot react fast enough upon request of the satellite computer. These data losses only occur because the satellite computer has to go on in his real time process.

Likewise, because of the small memory, it is impossible to run larger FORTRAN programs as for example library search and isotope identification for γ-ray spectroscopy or library search for mass spectroscopy. That is why we are planning to extend the memory of the central computer and to increase the storage capacity of the rotating memory, too. Furthermore, we are planning to install a second processor in the central computer, which shall be used as a communication processor for a connection to the large computer of the computer center of the University of Hamburg via a 48 K bit/s line.

In spite of the deficiencies just mentioned the satellite computer network is working very well. Table 2 gives the medium yield of data per day from the different instruments which have to be processed by the system. The central computer is in operation for the whole day. Since 1976 the availability of the system has been about 95 %. We had only very few serious interruptions.

TABLE 6.2 MEDIUM YIELD OF DATA PER DAY FROM
          THE DIFFERENT INSTRUMENTS

KIND OF DATA	DATA SETS PER DAY	LENGTH OF A DATA SET
Mass Spectra	30 - 300	1 K words
GC-MS Runs	2 - 6	
Gamma-Ray Spectra	2 - 20	8 K words
UV Spectra	100 - 1000	120 words

In the last years progress in computer technics has been great and parallel to that hardware costs diminished rapidly. Total costs for our computer network had been about 1 600 000.-- DM in 1973. So nowadays, one reason to form a computer network that is sharing hardware, has become far less important. On the other side, costs for the development of software, which is necessary for new computer networks like ours, have increased rapidly.

Many instruments nowadays are delivered with dedicated computers the capabilities of which are often sufficient even for scientific research. But if there are identical analytical instruments, using the same data base for library search or if measurements from different instruments (IR-, NMR- or mass spectrometer) are used for combined identification of e.g. chemical compounds, computer networks are still convenient.

ACKNOWLEDGEMENT

We thank the Deutsche Forschungsgemeinschaft for financial grants which rendered the development of the satellite computer network possible, and we thank Dr. D. Buescher and H. Ravens (Varian MAT) for helping to develop the software and hardware of the system.

REFERENCES

1. S. D. Perone, *Anal. Chem.*, 43, 1288 (1971).
2. E. Ziegler, D. Henneberg, and G. Schomburg, *Anal. Chem.*, 42, No. 9, 51A (1970).
3. R. E. Dessy, *Anal. Chim. Acta*, 103, 459 (1978).
4. H. Ch. Broecker and H. G. W. Mueller, *Chromatographia*, 7, 432, (1974).
5. K. Nagorny and K. Zietlow, *Chem. Ing. Techn.*, 44, 1201 (1972).
6. J. Hinsch and J. B. Mielck, *Pharm. Ind.*, 40, 971 (1978).
7. H. Timm and K. Zietlow, *Conference on Computer Based Analytical Chemistry, Portoroz/Yugoslavia* (1979).

7

THE UNIVERSITY OF UTAH NMR LABORATORY MINICOMPUTER NETWORK

D. W. Alderman, W. D. Hamill, Jr.,

C. L. Mayne, D. M. Grant

INTRODUCTION

A small minicomputer network has been implemented in the NIH Regional Resource in NMR at the University of Utah. Shown in Fig. 1 is a diagram of the network, which connects the minicomputers in four FT NMR spectrometers to a small hub minicomputer via low speed serial lines. The small minicomputer buffers data for a large minicomputer to which it is connected by a high speed parallel link. The function of the network is to support our research in high resolution NMR of liquids and solids by providing a centralized facility for storing, retrieving, transforming, displaying, phasing, and plotting data taken on all of our Fourier transform NMR spectrometers. In addition, the large minicomputer in the network meets all of our Fortran computation requirements. This chapter discusses the interconnections required to implement the network and presents the details of the hardware and software employed to achieve our objectives. Some of the factors which we believe are responsible for the success of the network, and plans for its future development are also discussed.

NETWORK COMPONENTS

The principal component of the network is the Digital Equipment Corporation (DEC) PDP 11/70 with high speed floating point processor, 512 K bytes of semiconductor error correcting memory, a 5 M byte disk, two 800 bits per inch, 75 inches per second, 9 track, ½ inch magnetic tape (mag tape) drives, a refreshed graphics display, a 1000 lines per minute electrostatic matrix type printer-plotter, a system terminal, and five user terminals. The PDP 11/70 runs DEC's mapped RSX11M multiprogramming real time operating system which supports excellent utilities including a number of text editors, a sophisticated macro assembler, a relocating and overlaying linker, a debugger, numerous file handling and backup utilities, and Fortran IV Plus, an optimizing

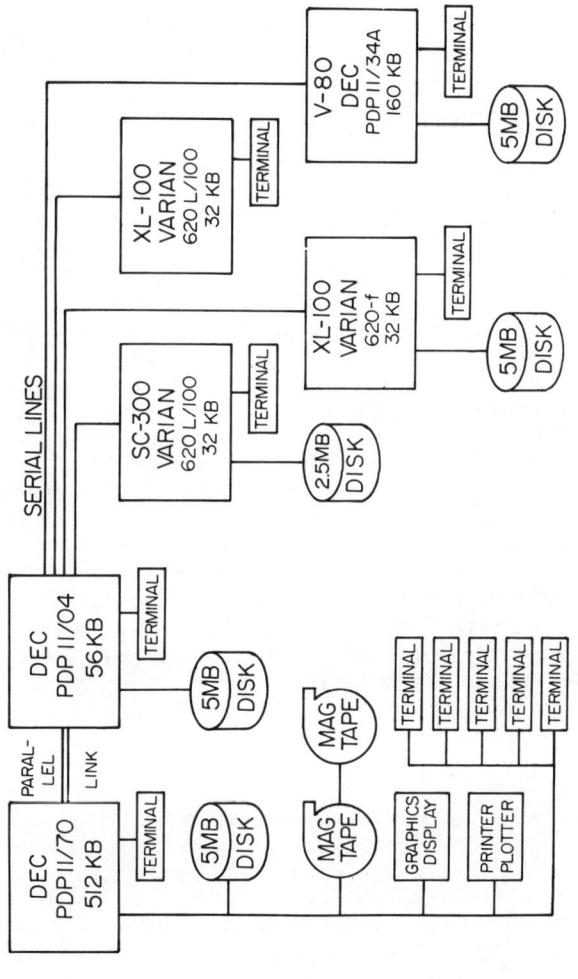

FIGURE 7.1 Schematic of Utah NMR lab computer network

Fortran compiler. The PDP 11/70 is connected via a high speed parallel link to a DEC PDP 11/04 which has 56 K bytes of semi-conductor memory with parity, a 5M byte disk, a KG11A cyclic redundancy polynomial checksum calculator, and a system terminal interface. It runs an unmapped RSX11M multiprogramming real time operating system which supports all of the utilities above except for Fortran IV Plus. The PDP 11/04 is connected to the four FT NMR spectrometers via low speed serial links.

The first of the spectrometers is a Varian SC-300 supercon-ducting instrument with a Varian 620 L/100 computer which has 32 K bytes of core memory and a 2.5 M byte disk. It runs the standard Varian SYMON disk FT software system which is built on the University of Oregon Chemistry Disk Operating System (CDOS). The second spectrometer is a Varian XL-100-15 FT instrument with a Varian 620 f computer which has 32 K bytes of core memory and a 5 M byte disk. This spectrometer runs the Varian FT16 program and also the Varian T1 research special program modified by us to interleave the taking of data at different recovery delays while storing all the accumulating data on the disk. These programs also run under the Oregon CDOS system. The third spectrometer is a Varian XL 100-15 FT instrument with a Varian 620 L/100 computer which has 32 K bytes of core memory but no disk. This spectro-meter runs only the Varian FT16 program. The last spectrometer, not yet fully integrated into the network, is an instrument of our own construction known as the ν-80 spectrometer which has a DEC PDP 11/34A computer with 160 K bytes of semiconductor memory with parity and a 5 M byte disk. This spectrometer uses a soft-ware system of our own design known as the Instrument Operating System (IOS) which runs under DEC's RT11 single user real time operating system.

NETWORK FUNCTION

The function of the network is arranged around the needs of the users of the four FT NMR spectrometers. Their first need is to store and retrieve data in the form of NMR free induction decays (FIDs) and programs and this is the first major function of the network. The network permits such data to be stored in and retrieved from standard named RSX11M files on any of three devices: the PDP 11/04 disk, the PDP 11/70 disk, and the PDP 11/70 mag tapes. This data has a number of sources. First, it may be a core image of any of the Varian computers. Second, it may be the contents of the data and parameter tables of the FT16 program. Third, it may be the contents of CDOS files on the disks of the two Varian computers so equipped. And fourth, it may be an image of the disk on the SC-300.

The above possibilities are used in a variety of ways. The core image of the FT16 program that runs on the diskless XL-100 is permanently stored on the PDP 11/04 disk so that is can easily be loaded at any time. The PDP 11/04 disk is also used for temporary storage of FID data from the same diskless XL-100 so that the data can be Fourier transformed and the spectrum examined without precluding further accumulation, as is the case in the absence of a mass storage device. The PDP 11/70 mag tapes are used extensively as repositories of data from all of the Varian spectrometers. Mag tape on our system stores roughly 10 M bytes per reel at a cost of $12. Such tapes are produced by transferring FT16 FID data and parameter tables from both XL-100's. Also, the CDOS file save capabilities are used for the disk equipped XL-100 when it runs the T1 program, and for the SC-300. Each user has a number of tapes on which he maintains data that he chooses to save. The SC-300 disk image capability is used for the special purpose of backing up on mag tape the original SC-300 software distribution disk.

With the exception of the initial startup of the network software on the PDP 11/04 and the PDP 11/70 and the mounting of mag tapes, all of the above services are available from the spectrometer terminal. That is, the user can store and retrieve data in any of the above ways by typing on any spectrometer terminal keyboard. From there he specifies the computer, either the PDP 11/04 or the PDP 11/70, the device, either disk or mag tape, and the name of the file to or from which data is to be moved. All users are trained to mount their own mag tapes and are obliged to do so. Since the largest distance between a spectrometer and the PDP 11/70 is only about 60 feet this is looked upon as a convenience rather than a burden.

The second major function of the network is to provide a sophisticated offline NMR spectrum plotting capability. The source of such spectra is FID data stored on the disk or mag tape of the PDP 11/70 in the manner described above. One of the user terminals is situated near the graphics display unit and the printer-plotter. The user accesses his data by device and file name through the terminal. The FID can then be weighted and Fourier transformed into a spectrum which is displayed on the graphics unit for phasing. Once this is done the spectrum can be plotted on the printer-plotter. Plots are not of the high quality obtained from an inking pen on a spectrometer plotter, but are produced much faster. Should a high quality plot be required the data can always be transferred back to the spectrometer and transformed and plotted there.

The third major function of the network, carried out exclusively on the PDP 11/70, is the Fortran computation done in support of our research program. Though this computation is done in an unconventional way which allows us to operate on more than 120,000 single precision floating point numbers in core, further details are outside the scope of this chapter.

NETWORK IMPLEMENTATION

The network communication hardware is implemented almost entirely with devices purchased from DEC and Varian. Because of the heterogeniety of the computers involved, however, it was not possible to use existing software in the network. Instead, it was necessary to devise a simple communication protocol and to integrate it into the software running on the various computers in the network. Further, supervisory programs were written to run on the PDP 11/04 and PDP 11/70 to accomplish the necessary file operations.

<u>Spectrometer to PDP 11/04 Serial Lines</u>. These lines are implemented using Varian Model 620-82 Universal Asynchronous Serial Controllers (UASC) and DEC DL11D Asynchronous Serial Line Interfaces. Both of these devices are designed to send and receive data in the standard asynchronous serial format first used for teletypes and now a nearly universal standard for data terminals. They are configured to run at 9600 baud and to send eight data bits, no parity bit, and one stop bit. This results in a transmission rate of 960 eight bit bytes per second. The EIA RS 232/C line voltage standard is used to transmit over Belden 8723 cable. The longest cable run is 100 feet.

The communication protocol used over these lines is of our own devising and was designed with the following criteria: Since it was necessary to implement the protocol on at least two dissimilar computers, it had to be as simple as possible. Also, since binary data is transmitted, the protocol must be transparent, i.e. it must function properly no matter what the nature of the data. These criteria led to the use of a block oriented protocol with the data block length fixed at 524 bytes. Since our communications environment is relatively error free it was decided to emphasize error detection, but not error correction. Thus, the protocol uses the sophisticated CRC-16 checksum but does not attempt to correct errors except by retransmission of an entire block.

Figure 2 is a schematic representation of the serial line protocol. One complete cycle of the protocol moves a block of 524 bytes of data from one computer to another. The logical design of the operation of the network assures that the direction of transmission is known to both computers before operation

Network Implementation 183

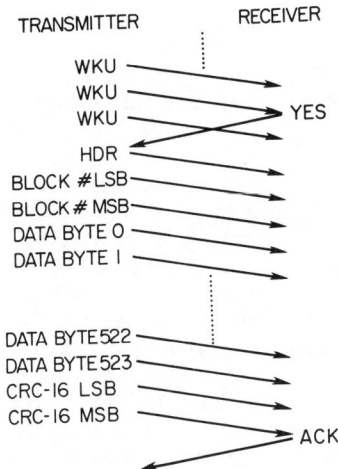

FIGURE 7.2 Schematic of serial line protocol.

starts, and thus no attempt is made to arbitrate in a situation where both computers want to transmit. From the outset one computer is designated the transmitter and the other the receiver. It was necessary to deal with the situation of lack of synchronization between the transmitter and the receiver, i.e. that the transmitter wants to send before the receiver is ready to listen, or vice versa. Each arrow in the figure represents the transmission of one eight bit byte over the line. There are five special control bytes: WKU for wake up; YES for ready to receive, HDR for data block header, ACK for positive acknowledgement, and NACK for negative acknowledgement.

From the point of view of the transmitter the protocol works as follows: The transmitter starts by clearing out any previously received bytes and then sending a WKU on the line. At the completion of the transmission of this WKU the transmitter looks to see whether it has received a byte. If it has and the byte is a YES the transmitter proceeds with the sending of HDR and the data as described below. If the received byte is not a YES an error exit is taken. If no byte has been received the transmitter sends another WKU and at the completion of the transmission again looks to see whether it has received a YES. This loop continues until some fixed number of WKU's have been sent at which point the transmitter concludes that the line is dead and takes an error exit. The number of WKU's sent divided by the byte rate is the length of time the transmitter is willing to try before giving up. This number can be tailored to different circumstances but is typically set at 1920 so that at 9600 baud the transmitter will try for about two seconds. When the received byte is a YES the transmitter concludes that the receiver is ready to receive data at the maximum rate. The transmitter then proceeds to send the HDR byte, the low and high bytes of a 16 bit block number, the 524 data bytes, and the low and high bytes of a 16 bit CRC-16 checksum calculated over the block number and 524 data bytes at the maximum transmission rate with no further interlock or acknowledgement. Once this is done the transmitter waits to receive a byte back. If the received byte is an ACK the transmitter concludes that the transmission has been successful and takes a normal completion exit. If the received byte is a NACK the protocol restarts at the point where it starts to send WKU's. This retry loop is taken three times before an error exit is taken. If the received byte is neither an ACK nor a NACK an error exit is taken.

From the point of view of the receiver the protocol works as follows: The receiver first clears out any previously received bytes and then passively waits for a byte to be received. If the received byte is a WKU the receiver does any housekeeping necessary to insure that it will be able to handle data at the full line speed, sends a YES, and waits for another byte. If the next received byte is a WKU it discards it and waits for another byte. If the next received byte is neither a WKU nor a HDR the receiver takes an error exit. If the next received byte is a HDR the receiver prepares to receive the block number, 524 bytes of data, and the two bytes of the CRC-16 checksum. Once these have been received it compares the received checksum with that which it has calculated and if they are the same the receiver sends an ACK and takes its normal completion exit. If the checksums are not the same the receiver sends a NACK and reinitializes itself to the point where it is waiting for WKU, unless this is the fourth NACK to be sent, in which case it takes an error exit.

A special short cycle mode of operation is incorporated into the protocol which is used to pass a special "end of file" message. In this mode the HDR byte, two block number bytes, 512 data bytes, and two checksum bytes are replaced by a single byte, a sixth control byte designated EOF. When the receiver gets EOF instead of HDR it immediately returns ACK and takes a special EOF exit.

On the Varian computers the protocol is implemented as a subroutine which retains control of the cpu, handling the UASC by testing for done and looping until the entire data block is sent and the protocol is complete. This is practical because of the single stream processing done on the Varian computers.

On the PDP 11/34A in the ν-80 spectrometer the protocol is implemented as an RT 11 device driver. It runs under full interrupt control in a way appropriate to the real time nature of the IOS software.

On the central PDP 11/04 which handles multiple serial lines the protocol is implemented as a single controller-multiple drive device driver running under interrupt control and using the full facilities of the RSX 11M operating system. This driver is unique in that it permits a call for input on any of the lines, and then returns the number of the line over which data is actually received when it completes. The supervisory program uses this mode to listen to all of the serial lines in order to supply services upon request to all of the spectrometers. A second unique feature is that when one line becomes active the driver locks out the other lines for the short time it takes to transfer one block in order to avoid having to service interrupts from more than one line at a time. At the baud rates currently used this is not strictly necessary, but was done in anticipation of higher baud rate operation in the future.

PDP 11/04 to PDP 11/70 Parallel Link. The parallel interface is built from a pair of DEC DR11A 16 bit parallel input - 16 bit parallel output interfaces. These interfaces are program control and not direct memory access units. The data lines are simply cross linked, i.e. the 16 output lines from one are connected to the 16 input lines of the other, and vice versa. Thus, data loaded into the output register on one computer may be read from the input register on the other. Transfers are synchronized by also cross coupling auxiliary control lines in the interfaces so that an interrupt can be generated in the receiving computer when data is loaded by the transmitting computer and an interrupt can be generated in the transmitting computer when this data is read by the receiving computer. The hardware that accomplishes this cross coupling, consisting of a pair of cards each with two integrated circuits on them, is the only hardware in the network that has been built by us.

The communication protocol that is used across the parallel interface is similar but not identical to that which is used in the serial lines. The WKU, YES, HDR, EOF, ACK, and NACK aspects of the protocol remain the same. The differences are that data is transmitted in units of 16 bit words rather than eight bit bytes, a 16 bit word count is sent at the beginning of the data block, which can be up to 32,767 words long, and the CRC-16 checksum is not used.

The parallel protocol is implemented in an RSX 11M device driver identical copies of which execute on both the PDP 11/04 and PDP 11/70.

Network Programs.   Two programs, NET04 which runs on the PDP
11/04, and NET70 which runs on the PDP 11/70, supervise the operation of the network. NET04 listens to all of the serial lines
by calling the driver in the mode in which it accepts input from
any of the lines. A spectrometer computer initiates activity by
requesting a file operation. It specifies a computer, PDP 11/04
or PDP 11/70, a device, disk or mag tape, a file name, and
whether it wants to read or write the file. If the request is
for a PDP 11/04 disk file operation NET04 opens or creates, reads
or writes and transmits or recieves the file contents to or from
the spectrometer. If the request is for a PDP 11/70 disk or mag
tape file operation NET04 passes the request through the parallel
link to NET70 which performs the file operation.

Plot Facility Program.   The plot facility program consists of
a translator module, a transform module, a display-phase module,
and a plot module, all of which are written mostly in Fortran.
The translator's task is to read FIDs and spectrometer parameters
in files which have been created by the various spectrometer
software systems, convert them to a standard format, and write
them to a working file. The transform module then needs only to
process data in the standard format of the working file where
FIDs are represented in single precision floating point format
(24 bit mantissa and 8 bit binary exponent). The Fortran Fourier
transform program transforms an 8K floating point real FID into
a 4K complex spectrum in about 2.0 seconds. The display-phase
module display selectable portions of the resultant spectrum on
the graphics unit and allows it to be interactively scaled and
phased. The plot module employs software supplied by the manufacturer of the printer-plotter to produce a plot of the spectrum.

OVERVIEW

   Even at its present level of development the network is successful. With about two man-years of effort, almost exclusively
for software development, a very useful system has been implemented and provided to a variety of users. The principal reason
for this level of success with such a modest expenditure of manpower is that our goals were simple and well defined. Because
the network is connected to only one type of computerized instrument, FT NMR spectrometers, there has been a homogeneity of purpose in our work which has avoided the insurmountable problem of
responding to the varied demands of many different types of users.
Further, because each spectrometer has its own minicomputer it
can operate largely independently of the network. Such a
"loosely coupled" system allows the communication links to be

slow, which in turn makes it possible to use inexpensive standard
interfaces supplied by the computer manufacturers, and to use
them in a completely standard way. The decrement in functionality associated with this low speed has been minimal. It takes
about 20 seconds to transfer an 8K point spectrometer data table
to a PDP 11/70 peripheral (not counting any tape wind time).
This is a tiny fraction of the time usually associated with obtaining NMR data. It is admitted that the functions chosen to
date do not require much speed, and as more demanding tasks are
called for it will be necessary to upgrade the speed of the
serial lines.

It might be noted that the spectrometers included in the network vary a great deal in age (2 to 11 years). In the near
future one of the two XL-100's will be replaced with a new superconducting instrument. Because of the standard nature of the communication interfaces and the loose coupling of the network, integrating new equipment into it will be relatively simple. This
flexibility is an important feature of the network.

Current efforts are directed towards making the plot facility
more interactive and able to handle more types of data, further
integrating the ν-80 spectrometer into the network, and supplying
on line services that will enable diskless spectrometers to do
interleaved T_1 measurements.

An important side benefit that has been realized is to get
the bulk of the network function onto computer equipment from
the mainstream of the minicomputer industry. In contrast to the
minicomputers supplied with commercial NMR spectrometers today,
the DEC PDP 11 is supported by excellent general purpose software systems. This capability was crucial to the rapid implementation of the hub function. Also, the hub computers are reliable and maintainable locally by the manufacturer or "third
party" maintenance companies. Further, add on memory and a wide
variety of peripherals from both DEC and other manufacturers are
available, typically at very competitive prices.

This work was supported by National Institutes of Health
Grant No. RR 00574.

8

A PRODUCTION LABORATORY IMPLEMENTATION OF DECNET

R. F. Coley

INTRODUCTION

    The application of computer networks in the chemical laboratory described in this chapter is that presently under development by the Nuclear Stations Division of Commonwealth Edison Company. Commonwealth Edison Company is a large, private electric utility company which services the northern third of the State of Illinois. Commonwealth Edison Company has three nuclear power stations in operation and three other nuclear power stations under construction.
    The operation of nuclear power stations is regulated by several federal, state and local agencies. Additional requirements arise from agreements with equipment and fuel vendors. Further requirements arise from internal quality control practices. All such requirements stipulate a number of chemical and radiochemical tests and measurements. The necessary tests and measurements apply to various station process streams and effluent paths.

<u>Analytical Facilities and Staff</u>. To perform required tests and measurements, each nuclear power station is equipped with at least two analytical laboratories and one radioactivity counting room. Analytical instrumentation varies in complexity from a simple conductivity bridge to gamma-ray spectrometry systems.
    In the Commonwealth Edison Company, specific tests and measurements are performed by unionized technicians. The technicians are combined radiation protection (health physics) and chemistry technicians. As many as 37 such technicians rotate through a variety of health physics, radiochemistry and chemistry tasks at the operating nuclear power stations. The work distribution is such that a given technician is assigned to the chemical laboratory for as little as two weeks in a nine month period. At the present time the technicians are not required to have a chemistry background. They are provided with job training specially designed by Commonwealth Edison and work under the

direct supervision of a laboratory foreman.

Procedures for tests and measurements are developed by the chemists who have functional control of the work of the technicians and foremen. Each operating station is staffed with a station chemist and two or three assistant chemists. The present requirement for assistant chemists is a B.S. degree with a major in chemistry or chemical engineering. The station chemist is required, in addition, to have one year experience in radiochemistry at an operating nuclear power station. Responsibility for the timeliness and correctness of analytical results lies with the station chemist.

THE TIME FOR ACTION

During the growth of Commonwealth Edison, three previously independent forces began to focus the need for a single action plan to meet analytical requirements within the nuclear division. These forces are briefly described in this section.

Modernization of Operating Stations. In the four year period from 1970 to 1974, Commonwealth Edison Company brought six nuclear reactor units on line -- two each at Dresden, Quad Cities and Zion nuclear power stations. Much of the analytical instrumentation for these stations had been selected by 1971. By 1979 the analytical instrumentation used to measure quantities of radioactive materials at the operating nuclear power stations was outdated and no longer supported by the vendor. Additional instrumentation was needed to meet increased Nuclear Regulatory Commission (NRC) surveillance requirements.

Equipping of New Stations. Commonwealth Edison was planning for six more reactors at three new sites -- LaSalle, Byron and Braidwood with then anticipated start-up dates between 1980 and 1983. Analytical equipment had to be selected for these new stations. Previous corporate decisions had dictated that the instrumentation for both Byron and Braidwood would be identical -- a step in a desired direction toward standardization.

Developing a Comprehensive Quality Control Program. Commonwealth Edison Company was in the process of establishing a comprehensive quality control program for the measurement of radioactive effluents. The quality control program was being modified to conform to the recently issued Nuclear Regulatory Commission (NRC) Regulatory Guide 4.15[1].

DEFINITION OF NEEDS

Faced with the replacement of equipment for three existing stations, the initial purchase of equipment for three new stations and the development of a comprehensive analytical quality control program, the staff of Commonwealth Edison's Central Radioanalytical Facility entered into a detailed study of analytical requirements.

Identified Concerns. A number of concerns were identified:

Concern 1 - Increased Sensitivity: Throughout the 1970's there had emerged requirements for increased analytical sensitivity. For example, primary feedwater metals had to be measured in the parts per billion range at boiling water reactor stations whereas parts per million analyses had been acceptable at fossil stations.

Concern 2 - Increased Regulations and Reporting Requirements: During the years from 1970 through 1979 regulatory agencies had greatly increased regulations and reporting requirements related to the measurements of quantities of radioactive materials in station process streams and station effluents. What had previously been satisfied by a gross radioactivity measurement now required a quantitative analysis for each identifiable radionuclide. High resolution gamma-ray spectrometry had become the routine analysis rather than the special analysis.

Concern 3 - Increased Complexity: As a company, Commonwealth Edison was experiencing an overall increase in complexity of chemistry related activities. There were soon to be six nuclear stations to consider rather than three; and the operating stations were accumulating a massive quantity of analytical results from required tests and measurements.

Concern 4 - Lack of Standardization: Rapid growth of the nuclear division of Commonwealth Edison Company in the early seventies had permitted three nuclear stations to start up without standardized analytical practices and policies.

Concern 5 - Lack of a Central Information Center: No one location existed where analytical results from all three operating stations could be reviewed or compared.

THE ACTION PLAN

By mid 1979 the staff of the Central Radioanalytical Facility of Commonwealth Edison Company had outlined the basic requirements of an action plan for addressing the recognized concerns:

1. The need for increased sensitivity would have to be met by employing modern analytical instrumentation.

2. The increased regulations and reporting requirements would have to be met by automation of report generation.

3. The overall increase in complexity would have to be met with standardization of methods and automation of correlation of results.

4. Lack of standardization could be eliminated by developing standard algorithms, procedures, reports and training.

5. The lack of a central information center could be removed by the establishment of a central data bank for analytical results.

In the action plan three words appeared repeatedly -- modernization, automation and standardization. The question remained as to whether or not it was cost effective to snychronize the modernization of operating stations with the purchase of instrumentation for new stations. If so, the desired standardization could become a reality; automation could be achieved by the judicious selection of equipment; and a comprehensive quality control program could emerge.

The major portion of the cost of analytical instrumentation at these stations was associated with the gamma-ray spectrometry systems. By this time, the vendor had been selected to provide these systems for the Byron and Braidwood sites. This vendor, EG & G ORTEC, utilized Digital Equipment Corporation (DEC) PDP-11 series computers as part of the data acquisition and analysis systems.

In August of 1979, therefore, Digital Equipment Corporation was retained to collaborate with the staff of Edison's Central Radioanalytical Facility in an evaluation of the existing situation and to propose a detailed course of action. This study resulted in a conceptual design document[2] which recommended increased automation and standardization of analytical practices at Edison's nuclear stations.

A Production Laboratory Implementation of DECNET

Late in 1979 corporate approval and funding was granted for a coordinated program for modernization, automation and standardization of nuclear generating station counting room and analytical laboratory instrumentation. This program became known as the Commonwealth Edison Company Analytical Instrumentation Project. The project budget called for equipment investments of $1,597,000.00 (not including instrumentation for LaSalle, Byron and Braidwood purchased as part of new plant construction budgets) and a software development expense of $720,000.00. Because the previously selected station instrumentation included DEC PDP-11 computers, DEC was chosen as the central equipment vendor and as the software development specialist.

COMMONWEALTH EDISON COMPANY ANALYTICAL INSTRUMENTATION PROJECT

<u>Overview</u>. The Commonwealth Edison Company Analytical Instrumentation Project is designed to modernize, standardize and automate the activities of the counting rooms and analytical laboratories at Edison's nuclear generating stations under the umbrella of a comprehensive quality control program. The overall project completion was initially targeted at the end of 1983 -- the then anticipated date for operation of six nuclear power stations in the Edison service system. The 1983 design objective involves the following physical facilities:

1. Nuclear station counting rooms and analytical laboratories located at six nuclear power stations -- Dresden, Quad Cities, Zion, LaSalle, Byron and Braidwood.

2. Counting room simulator located at the Central Radioanalytical Facility in Edison's Technical Center in Maywood, IL.

3. Data Analysis and Central Records Management System located at the Central Radioanalytical Facility.

4. Mobile analytical facility based at the Central Radioanalytical Facility.

<u>Objectives</u>. The 1983 design objective includes the following specific goals:

1. Replacement of obsolete analytical instrumentation at the three operating nuclear power stations.

2. Installation of new analytical instrumentation at the three future nuclear power stations.

3. Development of a comprehensive analytical quality control program.

To accomplish these three goals simultaneously, the Central Radioanalytical Facility staff defined the following additional goals:

1. Standardization of analytical instrumentation.

2. Standardization of analytical procedures.

3. Creation of a central data bank for results of required tests and measurements.

4. Development of a standardized training program for nuclear station chemists, foremen and technicians.

Project Requirements. To meet the objectives of the project, the following specific requirements would have to be included:

Nuclear Generating Station: Each nuclear generating station would have to be provided with the following:

1. Standardized, modern, intelligent analytical instrumentation.

2. Standardized algorithms and procedures for required tests and measurements -- unalterable by station personnel.

3. Standardized training for technicians and chemists.

4. Programming capability in a high-level language for local engineering applications.

5. Ready access to historical data and to data from other nuclear stations in the Edison system.

6. Redundant CPU's (as part of the defense in depth philosophy) to assure station availability.

7. Local hard copy report generation capability.

8. Validation of analytical results prior to recording.

9. Correlation of results of different tests and measurements conducted on a single sample.

10. Control and documentation of required testing schedules for both calibrations and analyses.

Central Development and Control Laboratory: The Central Radioanalytical Facility would have to provide the following capabilities:

1. Central data bank for results of required tests and measurements from all six nuclear stations.

2. Central report generator capable of producing required reports in a standardized format promptly after arrival of data from the stations.

3. Central standardized application program development capability in a non-production environment.

4. Central log for notification of out-of-specification analytical results from the nuclear stations.

Counting Room Simulator Laboratory: The Central Radioanalytical Facility would have to provide the following capabilities:

1. Physical facilities (instrumentation and laboratory) for standardized analytical procedure and application program development in a non-production environment.

2. Physical facilities for standardized analytical equipment, procedure and application program testing in a non-production environment.

3. Facility to develop and conduct standardized training in a non-production environment.

Some of these requirements would be met by the network aspect of the project; other requirements by purchased hardware and software; still others by specially designed system software.

Network Topology and Communications Hardware. The network is configured with a single PDP-11/70 RSX-11M DECnet node, called the host system, located at Edison's Central Radioanalytical Facility. Linked to the central node are seven satellite nodes -- six nuclear stations and the counting room simulator. Each

satellite node has a pair of PDP-11/34 RSX-11M DECnet processors.
The satellite processor pairs run in a prime/shadow configuration. All analytical instrumentation is interfaced to the prime processor through manual switches. The satellite nodes are capable of full operation with only one processor functional. Generally, the prime processor is in an active mode and the shadow is in a standby mode. This is not strictly true, however, because the shadow processor is used for certain files and other system functions. The status of the shadow processor is maintained such that upon loss of the prime processor the system could be brought back up to full functionality without data loss using the shadow processor. Local engineering application programs not associated with the network project can be developed using the shadow processor.

Each satellite node is equipped with two physical DECnet links to the central node. One link is a 1800 bps full duplex (FDX) leased line using Bell System 202T modems and C2 conditioning. The other is a 1200 bps dial (DDD) backup line using Bell System 212A modems. The links can be manually switched between the two satellite processors just as with the instrumentation interfaces. At each satellite node a third DECnet link is used to connect the two processors. In addition to the six leased lines, the central node is equipped with four 212A modems available to the satellites on a first-come-first-served basis. Figure 8.1 provides a diagrammatic representation of the network topology.

Purchased Hardware and Software. The standardized hardware and software for each satellite node are listed in Tables 8.1 and 8.2, respectively.

TABLE 8.1.   SATELLITE NODE STANDARDIZED HARDWARE

QUANTITY	DESCRIPTION
2	PDP-11/34A with 248K bytes of memory
2	Plessey 10M byte disk (5M byte removable)
2	Texas Instruments model 810 printer
2	ORTEC model 7010 MCA and interface
2	DZ-11A 8 line network link multiplexer
2	DZ-11A 8 line peripheral multiplexer
2	DZ-11B 8 line peripheral multiplexer
2	VT100 CRT terminal (without advanced video option)
8	20 milliamp serial switch

TABLE 8.1 (Continued)

QUANTITY	DESCRIPTION
16	RS232 EIA serial switch
1	Advanced Electronics Design model 3100P floppy disk

TABLE 8.2. SATELLITE NODE STANDARDIZED SOFTWARE

QUANTITY	DESCRIPTION
1	RSX-11M, version 3.2
1	FORTRAN IV, version 2.2
1	DECnet-11M, version 2.0
1	ORTEC 6516 DMOS[a] software
1	Utility for transfer from floppy to Plessey disk

[a] Data Management and Operating System.

The hardware and software for the central node are listed in Tables 8.3 and 8.4, respectively.

TABLE 8.3. CENTRAL NODE HARDWARE

QUANTITY	DESCRIPTION
1	PDP-11/70 with 512K bytes of memory
1	FP-11C floating point processor
2	RM03 67M byte disk
1	Plessey 10M byte disk (5M byte removable)
1	TE16 magnetic tape drive
2	DZ-11E 16 line multiplexer
3	LA120 180 character per second hardcopy terminal
1	VT100 terminal (with advanced video option)
2	VT100 terminal (without advanced video option)

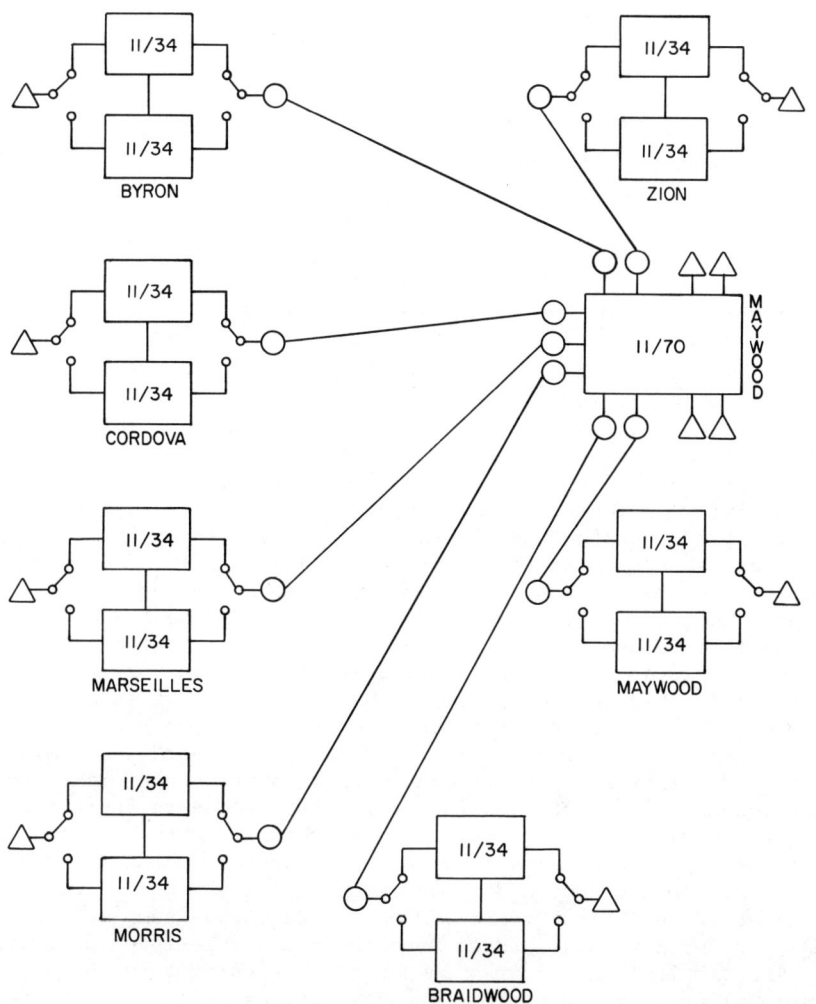

FIGURE 8.1. Network Topology for the Commonwealth Edison Company Analytical Instrumentation Project. The circles at each node represent the primary communications link using leased lines and Bell System 202T modems at 1800 bps. The triangles at each node represent the backup link using direct distance dial lines and Bell System 212A modems at 1200 bps.

TABLE 8.4.   CENTRAL NODE SOFTWARE

QUANTITY	DESCRIPTION
1	RSX-11M, version 3.2
1	RSX-11M autopatch
1	FORTRAN IV, version 2.2
1	DECnet-11M, version 2.0
1	SORT-11, version 2.0

Specially Designed System Software.   Each work shift at a nuclear power station has an associated schedule of analytical activities. A given shift schedule will specify a number of samples. Each sample, in turn, will require a number of analyses. For example, a given station might be required to do pH, conductivity, chloride and radionuclide analyses on a primary coolant sample during the day shift each day of the week.

The host system at the Central Radioanalytical Facility is used to specify the schedule of analyses to be run at the stations. The host system is also used to develop the standardized software to be used at the stations. Additionally, the host system serves as the central data analysis and records management system. In this last capacity the host system provides a central data bank and standardized central report generator.

The satellite nodes are designed to run standardized semi-automated analyses by interfacing with laboratory equipment and technicians. For each specific analytical procedure there exists a corresponding analysis-level program which prompts the technician for input, communicates with laboratory equipment, performs necessary computations and produces an analysis report. Because a number of analytical tests are performed on one sample, the analysis-level program passes results to a sample-level program.

The sample-level program compares the results from the analysis-level programs to various expected values or limiting values. The technician is notified when results are out of specification.

A schedule-of-analyses level program tracks the status of all analyses scheduled for a given shift -- to be run, running or completed. The schedule-level program also provides system command facility.

The hierarchy of the station application software is diagrammatically presented in Figure 8.2.

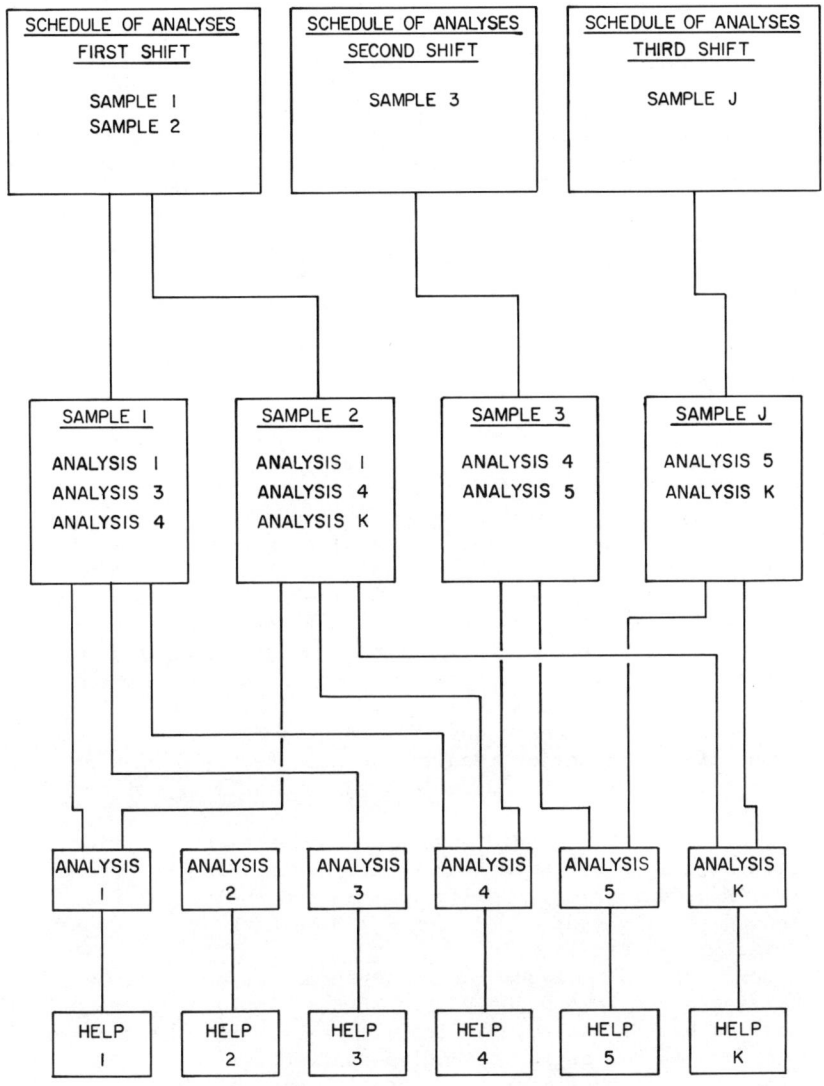

FIGURE 8.2. Hierarchy of specially designed Commonwealth Edison Company Automated Analytical Instrumentation System software.

System Operation. The following three sections describe the system operation from the perspective of the station technician.

Start of Work Shift: The technician using the system at a nuclear power station begins a work shift by logging onto the system. The system will indicate to the technician any instrument performance checks due for that shift. When completed, the performance check data will be verified against limits by a performance check program. If performance check results are out of specification, a chemist may override the check. Such an override will cause an entry into a log file and notification at the central log console. All performance check results are logged and transferred to the central data bank. When the log-on and performance check process is complete, all station terminals will display the names of all other samples to be run during that shift.

Start of Analysis: The technician begins an analysis by specifying a particular sample and analysis. The schedule-level program will then initiate the specified analysis-level program. The analysis-level program prompts for technician input. When the analysis is complete, the system will compare the results with existing limits. If the results fall outside the limits, the system will allow the operator to re-run the analysis or to call up a HELP file. (The HELP file contains text designed to help a technician recover from anticipated technique and procedural errors.) If the technician is convinced that the out-of-specification results are real, he must request that a chemist override the limit check. When the results are within limits or when out-of-specification results are accepted by a chemist, the analysis-level program produces a local analysis report and exits.

A record of all technician re-runs and chemist overrides is created locally and transmitted to the central log console at the Central Radioanalytical Facility. Selected analytical results are automatically sent to the central data bank at the Central Radioanalytical Facility.

Non-scheduled analyses can be performed at any time by specifying a sample and analysis. If the sample-analysis pair is recognized by the system, results will be checked and processed in the same manner as for scheduled analyses.

Analyses can be performed on special samples by specifying "special sample" as the sample name. Analyses of special samples produce local reports only -- no limit checking is performed and results are not transmitted to the central data bank.

Analyses can be aborted while in progress. After a scheduled analysis is aborted its status becomes "to be run". An aborted analysis results in an entry into the local and central log files.

End of Work Shift: The technician ends a work shift by logging off the system. If all analyses scheduled for that shift have been completed the system will complete the shift and return the station terminal displays to a pre-logon condition. If any analyses were not completed the system produces a list of incomplete samples and analyses.

A chemist is required to continue the log off process with incomplete analyses. The system asks the chemist for directions concerning incomplete analyses. Incomplete procedures can be overriden (eliminated from the schedule) or carried over (added to the schedule for the next shift). Alternatively, the log off process can be aborted so that the technician can continue until done. A record of analysis overrides is transmitted to the host system at the Central Radioanalytical Facility.

General Features. The system is capable of simultaneous use from any of four terminals -- the two MCA terminals and the two VT100 terminals. (An analysis started at one terminal must, however, be completed at the same terminal.) In this regard, the system is a multiuser, multitask system. That is, a technician can run an analysis at one terminal independent of another technician using another terminal.

Four system status displays are available to the technicians:

1. Display all samples which are currently running, that is, samples for which at least one analysis has begun.

2. Display all samples which are completed, that is, samples for which all analyses have been performed.

3. Display all samples to be run, that is, samples for which none of the analyses has been started.

4. Display the status of all analyses for a particular sample, this is, indicate "running", "completed" or "to be run" for all analyses associated with a particular sample.

Log files of chemist overrides, technican re-executions and program faults are created locally and at the central system. If the communications link to central is down, the local log file can be transmitted to central when communications have been re-established using the system store and forward capability. If the central log console is down, the central log file can be printed when the log console has been re-activated.

Local files are also created which contain the results which are to be transmitted to the central data bank so that data are not lost if the network link is down.

File transfers from central to the satellite nodes are accomplished using standard DECnet-11M utilities. Satellite to central transmissions of analytical results and log file information utilize the real time task-to-task (program-to-program) communications ability of DECnet-11M.

Special Operating Conditions. A chemist can identify certain special operating conditions, such as a reactor start up, to the system. The system will allow a chemist to add special sample-analysis sequences to the shift schedule. The system will also allow a chemist to delete analyses. All additions and deletions are recorded in the log files and transmitted to the central log console.

Under certain special operating conditions some analyses are required at periodic intervals of the order of hours. In such cases the system will prompt the technician at predetermined intervals until the analysis has been performed.

A chemist will inform the system when the special operating conditions end. At that time, any scheduled special operating conditions analyses will be cancelled. Central will be notified of the start and end of special operating conditions.

Manual Analyses. Because the system is complex, not all nuclear station analytical procedures will be converted from manual methods to semiautomatic methods by system installation time. In some cases this is not possible because of the lack of a suitable interface for existing instrumentation. In other cases the conversion from manual to semiautomatic analysis will take place during a system enhancement phase scheduled to follow the initial system installation phase.

To handle existing manual methods, the system includes an analysis-level program for manual analyses. This program prompts the technician for results of manual analyses. Utilization of this program results in the desired limit checking and central data bank updating for manual analyses.

Automated Analytical Instrumentation System (AAIS). The collection of standardized application software and specially designed system software has been named the Commonwealth Edison Company Automated Analytical Instrumentation System (AAIS).

The algorithms for computations associated with all analyses are developed by the staff of the Central Radioanalytical Facility in collaboration with a task force consisting of chemists from Edison's nuclear division. This group also develops the standardized written procedures associated with the individual analyses. Desired input and output from analysis-level, sample-level and schedule-level programs are also specified by this group.

Application software for radionuclide analyses utilizing minicomputer assisted gamma-ray spectrometry systems had been developed by the staff of Edison's Central Radioanalytical Facility for operation on now obsolete Hewlett-Packard equipment. Digital Equipment Corporation personnel were retained to convert the radionuclide application programs to code suitable for running under RSX-11M, to develop the analysis-level, sample-level and schedule-level software and to combine all software into a DECnet-11M working system -- the Commonwealth Edison AAIS.

IMPLEMENTATION

As stated at the outset, this system is under development at the present time. Although the overall project objectives are targeted for 1983, the Automated Analytical Instrumentation System (AAIS) is scheduled to be functional in early 1981. The overall Analytical Instrumentation Project is designed according to the following phased development and implementation schedule:

1980. The first major division of the project is called the 1980 product. This portion of the project, actually scheduled for completion in 1981, includes the acquisition and installation of instrumentation and the development and implementation of the original AAIS software.

Purchased Hardware and Software: The central node hardware and software listed in Tables 8.3 and 8.4 were purchased and installed by January of 1980. The standardized satellite node hardware and software listed in Tables 8.1 and 8.2 is purchased and installed on a station-by-station basis. Because of delays in the start-up schedules for LaSalle (now 1981 and 1982), Byron (now 1983 and 1984) and Braidwood (now 1985 and 1986), some instrumentation expenditures have been delayed. The present schedule for installation of purchased hardware and software is presented in Table 8.5.

TABLE 8.5.   HARDWARE AND SOFTWARE INSTALLATION SCHEDULE

NODE	HARDWARE	SOFTWARE
Central Radioanalytical Facility	JAN80	FEB81
Counting Room Simulator	JAN80	FEB81
Dresden	DEC80	MAY81
Quad Cities	JUN80	SEP81
Zion	DEC79	AUG81
LaSalle	SEP80	JUN81
Byron	NOV80	OCT81
Braidwood	DEC82[a]	-----[b]
Mobile Analytical Facility	DEC83[a]	-----[b]

[a]Not yet ordered.
[b]Not yet scheduled.

During the course of development, consideration was given to the purchase of DATATRIEVE -- a DEC software product specially designed for easy manipulation of data for tabulation and reporting purposes. It was decided not to attempt to utilize DATATRIEVE mainly because the product does not provide for data in E format (exponential notation) -- a serious limitation for scientific applications.

During the development of the 1980 product more personnel were slowly added to full time AAIS development work. To utilize the additional personnel efficiently more VT100 terminals were added to the PDP-11/70 system. By August of 1980 five additional VT100 terminals had been added and a strain on the system resources emerged. The system pool area (a common memory area accessed by various programs) started to become consumed at a regular frequency with resultant system crashes.

To alleviate the problem, a decision was made to upgrade the operating system to RSX-11M-PLUS which is designed to optimize the performance of the PDP-11/70. The size of the pool space was increased by a factor of two with the PLUS upgrade. The upgrade to RSX-11M-PLUS required a corresponding upgrade to DECnet-11M-PLUS version 1.0 (a DECnet phase III product) for compatibility at the central node. The satellite nodes continue to operate under RSX-11M and DECnet-11M version 2.0 (DECnet phase II).

Application Software and Standardized Procedures: The development portion of the 1980 product includes the preparation of the AAIS software and the associated standardized written procedures for twenty-two semiautomated analytical methods. These procedures involve communications with four types of analytical instrumentation -- top loading balance, pH meter, gas flow proportional counter and gamma-ray spectrometry system. Table 8.6 lists these procedures and the associated instrumentation.

TABLE 8.6. SEMIAUTOMATED ANALYTICAL PROCEDURES AND ASSOCIATED INSTRUMENTATION

ANALYTICAL PROCEDURE	INSTRUMENTATION
pH determination[a]	pH meter
Balance performance test	Balance
Gross alpha activity (liquid) Gross beta activity (liquid) Gross beta activity (vapor) Proportional counter efficiency Proportional counter performance test	Proportional Counter
General radionuclide analysis BWR[b] coolant radionuclide analysis PWR[c] coolant radionuclide analysis Liquid waste radionuclide analysis Dissolved gas radionuclide analysis	Balance and Gamma-ray Spectrometry System
Filterable solids Catalytic recombiner effluent Charcoal adsorber effluent Chimney effluent SJAE[d] effluent Waste gas Airborne particulate Radioiodine Ge(Li)[e] efficiency Ge(Li)[e] performance test	Gamma-ray Spectrometry System

[a]Includes pH meter calibration check.
[b]Boiling water reactor.
[c]Pressurized water reactor.
[d]Steam jet air ejector.
[e]Lithium-drifted germanium gamma-ray spectrometer.

The 1980 product also includes standardized analytical procedures for the thirty-one manual methods listed in Table 8.7. Because the technician enters analytical results into the system for manual procedures, these methods are independent of the instrumentation actually used for the analysis.

TABLE 8.7.   MANUAL ANALYTICAL PROCEDURES

Gross alpha activity	Insoluble iron
Ammonia	Iron-55
BOD (biochemical oxygen demand)	Metal analysis
Boron	Oil and grease
Cation conductivity	Phenol
Chloride	Phosphorus-32
Conductivity	Silica
Cyanide	Sodium pentaborate
Dissolved oxygen	Strontium-89
Fecal coliform	Strontium-90
Fluoride	TDS (total dissolved solids)
Free chlorine	Total residual chlorine
Free chlorine consumption	Tritium
Free hydroxide	TSS (total suspended solids)
Hydrazine	Turbidity
Hydrogen	

All analysis-level application software is written in FORTRAN with special program calls to receive system data, communicate with instruments and create files.

The present system design and physical capacity of the central data bank is such that results from sixty thousand analyses can be stored on one RM03 (five platter) disk pack. For six nuclear stations running approximately 100 analyses per day, this allows for the storage of 100 days results on one disk pack.

<u>Interface Software</u>: The analysis-level application programs do not directly access the instrumentation. Specially designed programs called instrument interfaces communicate with the instrumentation. Information is passed between analysis-level programs and instrument interfaces using program-to-program communication techniques. The use of separate instrument interface programs reduces the size of the analysis-level programs and provides independence from differences in instrumentation.

Instrument interface software has been developed by DEC personnel for the instrumentation listed in Table 8.8. The instrument interface programs are generally FORTRAN language programs.

TABLE 8.8.   STANDARDIZED SATELLITE NODE INSTRUMENTATION FOR WHICH SOFTWARE INTERFACES HAVE BEEN DEVELOPED

QUANTITY	DESCRIPTION
2	Sartorius model 1204P top loading balance with 1200 series BCD output board and model A 7053-04 RS232 converter
2	Beckman SelectIon 5000 Ion Analyzer (pH meter)
2	Canberra model 2201S automatic alpha/beta counting system with model 1788 communications interface
1	Canberra model 2201 automatic low level alpha/beta counting system with model 1788 communications interface

<u>System Software and Standardized Reports</u>: Under the general heading of system software is a multitude of programs which create, modify and read files, initialize and update system tables, control displays, control execution of other programs, control terminal access, validate results, control network transfers, log messages and perform miscellaneous clean up and other tasks specific to the AAIS.

As part of the 1980 product, two major reporting functions will be automated:

1. Station specific chemistry summary reports will be produced on request. These reports will tabularly summarize all analytical results for specified samples for a specified interval of time. These reports will be produced at the host. A copy can be sent over the communications network to the satellite. Presently, such chemistry logs are prepared by hand.

2. The monthly radioactive effluent report for Dresden, Quad Cities or Zion Station will be produced by specification of station and month desired. This multipage reporting requirement is presently met by manual methods.

In addition, a report editing utility will be part of the central system. This utility will provide the capability of manual restoration of the central data bank following a head crash. (Future plans call for addition of two RM03 disk drives at the central facility to act as a central data bank back up.) The utility will also provide for addition or removal of selected results from the data bank.

Manpower Expenditure: The anticipated manpower expenditures for the 1980 portion of this project are presented in Table 8.9.

TABLE 8.9.   MANPOWER EXPENDITURES FOR THE 1980 PORTION OF EDISON'S ANALYTICAL INSTRUMENTATION PROJECT

ITEM	MAN-HOURS
DEC Consultants	16500[a]
CECO Chemists Task Force	2700
CECO Central Radioanalytical Facility Staff	23900
Total	43100[b]

[a] Includes 160 hours for conceptual design.
[b] Approximately 21 man-years.

The tabulated manpower expenditures include full time consultants from Digital Equipment Corporation in residence at Edison's Central Radioanalytical Facility. The staff of Edison's Central Radioanalytical Facility has prime responsibility for the project and therefore has dedicated full time to the project.

1981-1983.   The 1981 to 1983 phase of the overall project is dedicated to enhancement of the 1980 product.

Purchased Hardware and Software: Original plans called for the counting room simulator to be equipped with a subset of the analytical instrumentation in use at the nuclear generating stations. Present plans now call for a complete simulator facility. Without a complete simulator, the development, testing and training efforts would suffer.

As mentioned earlier, additional RM03 disk drives are planned for central data bank back up. To improve the quality of report generation, a letter quality printer, word processing terminal and word processing software are being considered. The word processing subsystem would also be used for more efficient preparation of standardized written analytical procedures. Addition of plotting capability at the central node is also planned.

To improve the efficiency of development work and the quality of the associated documentation, present plans call for the addition of a line printer to the central PDP-11/70 system.

Because the Central Radioanalytical Facility is manned on the day shift only, troubleshooting support is presently not available to the stations on weekends and back shifts. To remove this deficiency present plans call for the addition of portable acoustically coupled terminals to the system. A staff specialist on call would have such a terminal at his residence for emergency purposes.

With expansion of the number of analytical procedures incorporated into the system more VT100 terminals (three or four) will have to be added at each nuclear station. An additional DZ11 eight line multiplexer and ten more RS232 EIA serial switches are planned to accommodate additional interfaced instrumentation at each satellite node.

With anticipated expansion of the system capability it is expected that the PDP-11/34 processors will become limiting. Thus it is expected that the processors will have to be upgraded at some time in the future.

Present plans do not call for the implementation of DECnet phase III at the stations (DECnet phase II provides point-to-point physical and logical connections. DECnet phase III products extend the physical connectivity with multi-point and the logical connectivity through routing.) Phase III would allow station-to-station forwarding of data and control elements. One of the most attractive features of DECnet phase III is the software upgrade service which is not available with DECnet phase II. Installation of DECnet phase III at the satellite nodes however would consume additional PDP-11/34 core space with resultant degradation in AAIS system performance. Present Edison needs do not warrant such an upgrade at the expense of performance.

It is planned that the mobile analytical facility will contain a subset of station analytical equipment. A communications link between the mobile facility and the network is planned -- perhaps using microware techniques.

Application Software and Standardized Procedures: In the 1980 product the previously developed radionuclide analysis application software was translated only. In 1981 and beyond, present plans call for these application programs to be enhanced by the addition of error analysis and other technical upgrades.

The present radionuclide library only provides for up to three gamma rays for each of 46 radionuclides. Present plans call for increasing the library to five gamma rays for each of 100 radionuclides. These analyses are also system limited to three spectrometers with up to 18 counting geometries each. Plans include expansion to 6 spectrometers of up to 60 counting geometries each.

Additional analytical methods will be converted from manual to semiautomatic. Present plans call for the inclusion of dissolved oxygen, chloride and fluoride analyses as SelectIon 5000 interfaced procedures.

Interface Software: Additional instrumentation will be interfaced to the system. Inclusion of Packard Instrument Company liquid scintillation counting equipment (models 460C/D and 2660) for which the hardware interfaces already exist is planned for 1981. Atomic absorption and UV-VIS spectrophotometers, analytical balances, conductivity instruments and a satellite plotting capability are being considered for addition to the system. The ultimate goal is to have all station analyses utilize instrumentation with digital output and interface capabilities.

System Software and Standardized Reports: The satellite nodes are being equipped with a portable gamma-ray spectrometry system. It is desired to be able to process spectrum files from the portable system using the AAIS software. Presently the AAIS contains a facility for analyzing gamma-ray spectra from a spectrum file rather than from the gamma-ray spectrometry system memory. To include the ability to handle spectra from the portable system requires the development of a utility to re-format the spectrum files from the portable system.

Certain other system software enhancements are also planned. It is desired to have the ability to display at the central node a copy of certain displays which are presently only available at the satellite nodes. It is also planned to increase the satellite node display options. It seems desirable, for example, to allow the technician to view the HELP file for an analysis prior to running the analysis. At present the HELP file is only available after the technician has obtained out-of-specification results.

As mentioned earlier, digital plotting capabilities are under investigation. Station personnel presently manually prepare several plots as control methods for trending of various parameters. Such plots cover various time intervals from months to entire fuel cycles and even years. Present plans call for converting a general plotting program (previously developed by the staff of the Central Radioanalytical Facility) for use with the AAIS. A central version would incorporate a four pen plotter and be used for plotting data stored in the central data bank. A satellite version would use a single pen plotter for producing local hard copy for data trending.

Additional standardized reports will be formulated. The planned reports are listed in Table 8.10.

TABLE 8.10.   PLANNED STANDARDIZED REPORTS

NRC Semiannual Radioactive Effluents Report
BWR[a] Fuel Warranty Chemistry Report
Monthly Illinois NPDES[b] Report
Quarterly Federal NPDES[b] Report
PWR[c] Chemistry Report
US NRC Confirmatory Measurements Program Report
CECO Confirmatory Measurements Program Report

---

[a] Boiling Water Reactor.
[b] National Pollution Discharge Elimination System.
[c] Pressurized Water Reactor.

Manpower Expenditures: It is anticipated that the Analytical Instrumentation Project will continue to be the primary responsibility of Edison's Central Radioanalytical Facility. As such the project will demand another forty to fifty man-years of effort in the next three years, especially if the plans to develop a standardized training program are to become reality. In addition, at least one full time resident consultant (DEC) will be required for system management and the transition of expertise from DEC to CECO.

CONCLUSION

The Commonwealth Edison Company Analytical Instrumentation Project is an innovative application of modern computer network techniques. The project is complex and involves considerable expenditures of capital and manpower. The Automated Analytical Instrumentation System, when in operation, will provide the following improvements in quality control over the present manual methods:

1. On a work shift basis, the system schedules procedures to be run at the stations and notifies the central facility when procedures are not run.

2. The system assists the technician in performing the analytical procedures by prompting the technician for input and obtaining results from the instruments.

3. The system performs required computations (according to standardized algorithms), thereby eliminating computational errors.

4. The system validates the results of analyses depending on the sample type, thereby eliminating the chance of recording unreal out-of-specification results.

5. The system produces standardized local reports of analysis results, thereby eliminating transcription errors.

6. The system sends results to the central site, thereby producing a central data bank.

7. The system produces required reports at the central site in a standardized format, thereby reducing transcription time and errors.

With future planned enhancements, the AAIS will provide the following additional improvements in quality control:

1. The system will provide standardized station plots of analytical results as a function of time for trending purposes.

2. The system will provide central plots of various parameters from the central data bank versus time for trending and station-to-station comparison purposes.

In addition, the overall project will provide standardized training for chemistry personnel.

The overall philosopy is obvious -- it is cost effective to perform required analytical tests and measurements at six nuclear generating stations with standardized methods and instrumentation using personnel who have been provided with standardized training. Under such a program, the probability of increasing the overall availability of nuclear generating stations is a reality. At present, a two unit nuclear generating station produces approximately $1,000,000.00 in revenue for each day of operation. If improved analytical quality control results in as little as a few days of additional availability per reactor per year, the project expenditures are rapidly recovered. The staff of Edison's Central Radioanalytical Facility and Edison's corporate management are convinced that the Commonwealth Edison Company Analytical Instrumentation Project will be cost effective resulting in a net savings to their customers.

REFERENCES

1. *Quality Assurance for Radiological Monitoring Programs (Normal Operations) -- Effluent Streams and the Environment,* U. S. Nuclear Regulatory Commission Regulatory Guide 4.15, U. S. Nuclear Regulatory Commission, Washington, D. C. 20555 (February 1979).

2. J. F. Stach, et al., *Conceptual Design for the Standardization & Integration of Counting Room & Analytical Laboratory Instrumentation for Nuclear Generating Stations in a Cooperative & Co-Equal Distributed Network,* Commonwealth Edison Company, Chicago, IL 60690 (24AUG79).

INDEX

Absorption spectroscopy, 73
ADC, 8, 41, 66, 121ff
   successive approximation, 32
Advanced operating system
   (AOS), 145
ALGOL, 3
ALOHA, 16
Analog multiplexing, 32
Analog output channels, 33
Analog-to-digital converters,
   see ADC
Analytical electron
   microscopy, 73
Architectures, computer, 2ff
ARPANET, 18
Array processors, 27
ASCII, 4
Assemblers, 36, 37, 44
Asynchronous serial line, 34,
   168, 182
   RS232, 34
Auger electron spectrometer,
   39, 87
Autosamplers, 39

Background subtraction, 92
Backplane bus, see Bus
Backus Naur Form, 60
BASIC, 38, 45, 75, 174
Bell system modems, 195-196
Black box approach, 112
Block communication protocol,
   182
Bootstrap, 25, 31

Breakpoints, 38
Bus (computer)
   backplane bus, 28
   GPIB, 20, 104, 105
   HPIB, 20
   IEEE-488, 68, 106
   IEEE-696, 118
   multibus, 9, 78
   S-100, 9, 118ff
Butterworth filters, 121
Byte, 166

Cable length, 36
Cambridge ring, 21
Computer Aided Research
   Equipment Group (CARE), 41
Cassettes, 35
CDOS, Disk operating system
   for Varian 620 computer,
   179
Central computer, 162, 167
Checksum, 67
   cyclic redundancy check, 182
Chromatography, 39
Circular bus topology, see
   network
CODE definition, 46
Collisions, 18
COLON definition, 46
Colon processor, see Inner
   interpreter
Command substitution system
   (CSS), 82
Common-mode rejection, 33
Commonwealth Edison Company
   (CECO), 188, 190
   analytical facilities, 188

Index

analytical instrumentation project, 192, 203, 211, 213
automated analytical instrumentation system (AAIS), 199, 202, 203, 204, 205, 207, 210, 211
communications hardware, 194, 195, 196, 209
data analysis and central records management system, 192, 197
interface software, 206, 207, 210
mobile analytical facility, 192, 204, 209
network topology, 194, 195, 196
system operation, 200
technical center, 192, 196
Communication, bit-serial, 36
asynchronous, 26
computer-computer, 31
current loop, 36
full-duplex, 36
loop control, 26
parallel digital, 35
RS232, 36
synchronous, 26
Communication line, 162, 167, 169
Communication protocol, 167, 169
asynchronous serial, 182
block, 182
parallel data, 185
Communication system, ring-organized, 36, see also Network
Compilers, 37, 44
Computational capacity, 28
Computer bus, see Bus
Control, I/O sequencer, 33
structures, 58
timing, 33
word, 169

Conversational Command Interpreter, 139
CRC-16, cyclic redundancy check, 182
CRD Honeywell DPS-2 computer, 75
Cross-assembly, 37
Cross coupling, parallel data link, 185
CRT graphic display, 30, 110-111, 137-138
Current loop, 36
Cyclic redundancy check, 180, 182
Cyclic voltammetry, 39

DAC, 33, 41
Daisy chain, 27
Data arrays, 56, 124
Data acquisition, 25
subsystem, 76, 119
Data base management, 60, 82-83
Data cartridge, 35
integrity, 30
logging, 39
rate, 169
structure, 45
transfer, 169, 171
Debug monitor, 38, 110, 139
DECNET, 20, 41, 188, 194, 195, 203-204, 209, also see Digital Equipment Corporation
Dedicated computer, 161, 175
Definitions, 57
Densitometer scan, 88
Design philosophy, 153
Device driver, RSX 11M operating system, 185
RT 11 operating system, 185
Dial-up modems, 39
Dictionary, 43, 45, 57
Differential scanning calorimetry, 80
Differential transceivers, 106
Diffraction patterns, 96

Index

Digital Equipment Corporation
 (DEC), 191, 192, 203, 206,
 208, 211
DATATRIEVE, 204
DL11D asynchronous serial
 interface, 182
DR11A parallel interface,
 185
DZ-11 multiplexers, 195,
 197, 209
FORTRAN IV, 197, 198, 206
FP-11C floating point
 processor, 197
LA120 hardcopy terminal, 197
LSI-11, 41
PDP-11 series computers,
 178, 180, 191, 192
PDP-11/34, 195, 196, 209
PDP-11/70, 194, 196, 197,
 204, 209
RM03 disk, 197, 206, 208
RSX-11M, 194, 195, 197, 198,
 203, 204
RSX-11M-PLUS, 204
SORT-11, 198
TE16 magnetic tape drive,
 197
VT100 CRT terminals, 195,
 197, 201, 204, 209
Direct memory access (DMA),
 32, 63, 64, 128, 145
Directory-driven random access
 storage, 35
Disk operating system, see
 Operating system
Disks, floppy, 10, 34
Display, mixed graphic and
 alphanumeric, 30, 137,
 146, 155
DMA sequencer, 128,
 see also Direct Memory
 Access
DMS, see Operating systems
Documentation, 40
Double potential step,
 chronoamperometry, 39
Down-loader, 37

Drive, cartridge, 35
 floppy disk, 34
Dummy tasks, 66
Dynamic memory, 29, 136

Editor, 44
Editors, text, 37
EIA RS232C, serial line
 voltage standard, 182, see
 also RS232
Electrochemical techniques,
 FT, 39
Electron microprobe, 73, 80
 diffusion profile, 85
Electrophoresis gels, 39
Elemental analysis, 73
Error control, 36
Error correction,
 retransmission, 182
Error detection, cyclic
 redundancy check, 182
ESCA spectrometer, 39
Ethernet, 21, 106, 159

Fetch, 56
FID, nuclear magnetic
 resonance free induction
 decay data, 180
File handling, 83, 84, 155ff
 maintenance, 37
 security, 74, 75, 153
First generation language, 43
Floating-point arithmetic, 36,
 38, 156
Floppy disk, 10, 34
Fluorescence life-time
 measurements, 39
Format, synchronous data
 encoding, 35
Formatters, 38
Formatting electronics, 35
FORTRAN, 3, 73, 75, 154
FORTRAN computation, 181
Fourier transform, ion
 cyclotron resonance, 35
 nmr spectrometers, 35, 163,
 178
 free induction decays, 180

FSUNMR, 148, 151
  baseline flattening, 157
  calculate relaxation, 157
  peak analysis, 156
  relaxation analysis, 157
FSUNMR utilities, 148, 150, 151, 155
FT, see Fourier transform
Full-duplex, 36

Gamma-ray spectrometer, 162, 164
Gamma-ray spectrometry system, 190, 191, 203, 205, 210
Gas chromatograph, 162
Gas chromatography, 36
Gas chromatography-Mass spectrometry, 73
GE laboratory automation system (LAS), 71
Gel permeation chromatogram, 87
GPIB, see Bus
Graphic image manipulation, 38
Graphics, 34
  nuclear magnetic resonance data, 181
Graphics displays, 98, 137, 145
Group code recording, 34

Hardware averager, 32,
  see also Signal averager
Header, 171
Hierarchial environment, 25
High-resolution graphics, 27, 137, 145
Host computer, 25
  polling, 36
Host support system, 30
HPIB, see Bus
HPLC, 39
Hub minicomputer, 178

Identification block, 173
IEEE-488, see Bus
IEEE-696, see Bus
Image, see Operating systems

Image buffer, 30
Indexed access method, 56
Infix notation, 53
Information block, 169
  system, 172
INFOS, see Operating systems
Infrared spectrometry, 80
  spectrum, 86
Initial program entry, 31
Input/output structure, 28
Instruction counter, 49
Instrument operating system (IOS), 180
Intel, 8080A microprocessor, 9, 31, 78
  8253 programmable interval timer, 124
  8257-5 DMA controller, 130
  8291 GPIB talker/listener, 110, 114
Intelligent interface, 112
Interactive compiler, 45
  interpreter, 45
Interdata 70, 73
Interface, asynchronous serial, 36
Interfacing, 25, 112, 134ff
Interpreters, 37, 44
  inner, 49
  number, 49
  outer, 46
I/O formatters, 36
IOS, instrument operating system, 180
IR-spectrometer, 163

Jump tables, 62

Kernel, 51, 61
Key word, 171

LAM (Laboratory Automation Module), 72, 76, 78
Layers, network, 11ff
LC GPC (Liquid and Gel Permeation Chromatography), 79

Libraries, memory image, 37
  relocatable, 37
Lines, asynchronous serial, 182
Linker/loader, 37
Liquid chromatogram, 89
Loaders, 37
Local area network
  (description), 19ff

Macrocode, 50
Macro-level utilities, 36
Magnetic tape, 180
Mainframes, 26
Mass spectrometry, 80, 85, 162
Master program, 171
Memory refresh, 29
Metallography and light
  microscopy, 73
Microcode, 50
Microcomputer, 25, 118ff
  history, 8ff
Microprocessor, 8, 25
Minicomputers, history, 6ff
Mnemonic, 171, 140ff
Modems, 26, 195-196
Molecular orbitals, 97
Mostek 14007 serial control
  unit, 134
MTM, see Multi-terminal
  monitor, 77, 82
Multibus, 9, 78
  see also Bus
Multichannel analyzer, 163
Multicomponent analysis, 39
Multiple-precision arithmetic, 36
Multiplexor, 32, 165
Multistream batch, 26
Multitasking, 63
Multi-terminal Monitor (MTM), 77, 82
Multiuser, 63
  system, 74

Network, broadcast, 16
  circular bus topology, 36
  hierarchial, 36
  long haul, 16ff
  local, 19ff
  mesh, 15ff
  microcomputer, 102ff
  minicomputer, 178
  multilaboratory computer, 36
  protocol, 67
  ring, 15
  star, 14
  tree, 15
  topologies, 14ff
Neutron activation analysis, 164, 174
Neutron generator, 163, 174
NMR, 119
  see also Nuclear magnetic
    resonance
Nodes, 14ff
Nuclear magnetic resonance,
  73, 79, 103, 118ff, 178
  spectrometers, 178
  spectroscopy, 178
Nuclear Regulatory Commission
  (NRC), 189, 211, 213
Numerical analysis, 39

Off-loading, 27
Operating system library, 170
Operating systems, 82, 153, 169
  AOS, 146
  CDOS
    for Varian 620 computer, 180
  CP/M, 10
  CP/Net, 10
  DMS, 60
  IMAGE, 60
  INFOS, 60
  MP/M, 10
  RSX11M for Digital Equipment
    Corporation PDP11
    computer, 178
  RT11 for Digital Equipment
    Corporation PDP11
    computer, 180
  SATKOS, 169
  VORTEX, 169

Opto-coupler, 166, 168
Opto-isolators, 36, 72
Organic separations, 73

Packets, 36
Packet switching, 18
Parallel data link, 185
Party-line cable, 36
PDP-11, see Digital Equipment Corporation
Peak detection, 39
Peak search, 93
Peripherals, 35
Perkin Elmer 8/32 computer, 77
Phasing, nuclear magnetic resonance spectra, 181
Plessey disk, 195, 197
Plotter, XY, 33
  digital, 77, 148ff
Plotting, nuclear magnetic resonance spectra, 148-150, 181
Polling transaction, 36
Position independent code, 62
Postfix notation, 53
Powder diffractometer scan, 86
Printers, 34, 77, 195
Printer/plotter, 26
Procedure-oriented language, 69
Program counter, 50
Program library, 169
Programmable asynchronous lines (PAL), 77
Programmable timer/pulse controller, 125ff
Program-to-program communication, 171
Protocol, asynchronous serial communications, 182
  parallel data link, 185
  transparent, 182
Pulse height ADC, 164, see also ADC

Quadrature data, 39, 119

Rabbit system, 163, 173
Random access memory, 25, 32
Raster-scan videographics, 137, 145
Real time experimental control, 25
Real time process, 162, 175
Recorder, audio cassette, 35
  analog XY, 34,
  see also Plotter
Reentrant code, 49
Relocatable code, 44
Round-robin, 65
RS-232, 36, 41, 68, 73, 197, 209
RS232C, EIA serial line voltage standard, 182
RS-422, 106

Sample and hold, 121
Satellite computer, 162, 168
Scanning electron microscopy, 73
Search systems, spectral, 40
Second generation, 44
SEM (see scanning electron microscope), 73, 74
Sequencer, high speed I/O, 32
Sequential programming, 33
Serial lines, asynchronous, 185
Shaft encoders, 72
Shared-resource computing, 35
Signal averager, 123
Single photon counting, 39
Single-step instruction execution, 38
Slave program, 171
Smoothing of data, 92
SNA, 68
Software architecture, 169
Software development, 36
Source text editor, 37
SPECPLOT, 84, 94, 98
  interactive cursor commands, 99
Spectra, manipulation, 156-157
Spooling, 146, 156

Stack, 49, 51
  maps, 52
  operators, 51
Staircase voltammetry, 39
Stepper motor, 39
Store, 56
STOIC, 53
Strain gauges, 166
Structured language, 58
Structured programming, 58
Subtraction of spectra, 91
Successive approximation ADC, 32
Surface analysis, 73, 80
System interconnection, 81
  see also Interfacing

Tabletting machine, 163, 166
Task builder, 66
Terminals, teleprinter, 34
  see also Printer
Text handlers, 36
Three-wire handshake, see "GPIB"
Third generation, 44
Time-sharing, 26
TLC plates, 39
Track/hold, 32
  see also Sample and hold
Transactions, computer/instrument, 29
Transient digitizer, 39
Transmission Electron Microscopy (TEM), 81, 96

Universal serial controller/monitor, 120, 133, 182
Utilities, assembly-language, 37
  library maintenance, 38
UV-spectrometers, 162, 166

Vector processors, 5, 27
Video display, 30, 33
  see also Graphics Displays
Video monitors, 30
Virtual arrays, 62

Virtual circuits, 110
Virtual computer emulator, 50
Virtual memory, 27, 61, 136, 151-152
Voltammetry, 39

WARPATH, 102
  protocols, 115
  topology, 109
WARPATH CHIEF, 110
Wire-and, 106

X.25, 68
X-ray diffraction, 73, 79
X-ray fluorescence spectroscopy, 73, 80

Zilog, 9
Z-80 microprocessor, 9, 110, 112, 118, 145

RAYMOND H. FOGLER LIBRARY
DATE DUE

OOKS ARE SUBJECT TO